T0252478

Tidy Modeling with R
A Framework for Modeling in the Tidyverse

Max Kuhn and Julia Silge

Beijing · Boston · Farnham · Sebastopol · Tokyo

Tidy Modeling with R

by Max Kuhn and Julia Silge

Copyright © 2022 Max Kuhn and Julia Silge. All rights reserved.

Published by O'Reilly Media, Inc., 1005 Gravenstein Highway North, Sebastopol, CA 95472.

O'Reilly books may be purchased for educational, business, or sales promotional use. Online editions are also available for most titles (*http://oreilly.com*). For more information, contact our corporate/institutional sales department: 800-998-9938 or *corporate@oreilly.com*.

Acquisitions Editor: Michelle Smith
Development Editor: Rita Fernando
Production Editor: Beth Kelly
Copyeditor: Piper Editorial Consulting, LLC
Proofreader: Tom Sullivan

Indexer: Potomac Indexing, LLC
Interior Designer: David Futato
Cover Designer: Karen Montgomery
Illustrator: Kate Dullea

July 2022: First Edition

Revision History for the First Edition
2022-07-12: First Release

See *http://oreilly.com/catalog/errata.csp?isbn=9781492096481* for release details.

The O'Reilly logo is a registered trademark of O'Reilly Media, Inc. *Tidy Modeling with R*, the cover image, and related trade dress are trademarks of O'Reilly Media, Inc.

The views expressed in this work are those of the authors and do not represent the publisher's views. While the publisher and the authors have used good faith efforts to ensure that the information and instructions contained in this work are accurate, the publisher and the authors disclaim all responsibility for errors or omissions, including without limitation responsibility for damages resulting from the use of or reliance on this work. Use of the information and instructions contained in this work is at your own risk. If any code samples or other technology this work contains or describes is subject to open source licenses or the intellectual property rights of others, it is your responsibility to ensure that your use thereof complies with such licenses and/or rights.

978-1-492-09648-1

[LSI]

To Amy: When you read this, know that I love you more today than every day before.
—M.K.

To Robert: Happy 20 years of choosing each other.
—J.S.

Table of Contents

Part III.   Tools for Creating Effective Models

Part IV.   Beyond the Basics

Preface

Welcome to *Tidy Modeling with R*! This book is a guide to using a collection of software in the R programming language for model building called tidymodels, and it has two main goals:

- First and foremost, this book provides a practical introduction to *how to use* these specific R packages to create models. We focus on a dialect of R called the tidyverse (*https://oreil.ly/xnx26*) that is designed with a consistent, human-centered philosophy and demonstrate how the tidyverse and the tidymodels packages can be used to produce high quality statistical and machine learning models.

- Second, this book will show you how to *develop good methodology and statistical practices*. Whenever possible, our software, documentation, and other materials attempt to prevent common pitfalls.

In Chapter 1, we outline a taxonomy for models and highlight what good software for modeling is like. The ideas and syntax of the tidyverse, which we introduce (or review) in Chapter 2, are the basis for the tidymodels approach to these challenges of methodology and practice. Chapter 3 provides a quick tour of conventional base R modeling functions and summarizes the unmet needs in that area.

After that, this book is separated into parts, starting with the basics of modeling with tidy data principles. Chapters 4–9 introduce an example data set on house prices and demonstrate how to use the fundamental tidymodels packages: recipes, parsnip, workflows, yardstick, and others.

The next part of the book moves forward with more details on the process of creating an effective model. Chapters 10–15 focus on creating good estimates of performance as well as tuning model hyperparameters.

Finally, the last section of this book, Chapters 16–21 cover other important topics for model building. We discuss more advanced feature engineering approaches like dimensionality reduction and encoding high-cardinality predictors, as well as how to answer questions about why a model makes certain predictions and when to trust your model predictions.

We do not assume that readers have extensive experience in model building and statistics. Some statistical knowledge is required, such as random sampling, variance, correlation, basic linear regression, and other topics that are usually found in a basic undergraduate statistics or data analysis course. We do assume that the reader is at least slightly familiar with dplyr, ggplot2, and the %>% "pipe" operator in R, and is interested in applying these tools to modeling. For users who don't yet have this background R knowledge, we recommend books such as *R for Data Science* (*https://r4ds.had.co.nz*) by Wickham and Grolemund (2016). Investigating and analyzing data is an important part of any model process.

This book is not intended to be a comprehensive reference on modeling techniques; we suggest other resources to learn more about the statistical methods themselves. For general background on the most common type of model, the linear model, we suggest Fox (2008). For predictive models, Kuhn and Johnson (2013) and Kuhn and Johnson (2020) are good resources. For machine learning methods, Goodfellow, Bengio, and Courville (2016) is an excellent (but formal) source of information. In some cases, we do describe the models we use in some detail, but in a way that is less mathematical, and hopefully more intuitive.

Conventions Used in This Book

The following typographical conventions are used in this book:

Italic

Indicates new terms, URLs, email addresses, filenames, and file extensions.

`Constant width`

Used for program listings, as well as within paragraphs to refer to program elements such as variable or function names, databases, data types, environment variables, statements, and keywords.

`Constant width bold`

Shows commands or other text that should be typed literally by the user.

`Constant width italic`

Shows text that should be replaced with user-supplied values or by values determined by context.

This element signifies a tip or suggestion.

 This element signifies a general note.

 This element indicates a warning or caution.

Using Code Examples

Supplemental material (code examples, exercises, etc.) is available for download at *https://github.com/tidymodels/TMwR*. This book was written with RStudio (*https://oreil.ly/bcWV6*) using bookdown (*http://bookdown.org*) (Xie 2016). We generated all plots in this book using ggplot2 (*https://oreil.ly/vEJBy*) and its black and white theme (theme_bw()). An online version (*https://tmwr.org*) of this book is available and will continue to evolve after publication of the physical book.

If you have a technical question or a problem using the code examples, please email to *bookquestions@oreilly.com*.

This book is here to help you get your job done. In general, if example code is offered with this book, you may use it in your programs and documentation. You do not need to contact us for permission unless you're reproducing a significant portion of the code. For example, writing a program that uses several chunks of code from this book does not require permission. Selling or distributing examples from O'Reilly books does require permission. Answering a question by citing this book and quoting example code does not require permission. Incorporating a significant amount of example code from this book into your product's documentation does require permission.

We appreciate, but generally do not require, attribution. An attribution usually includes the title, author, publisher, and ISBN. For example: "*Tidy Modeling with R* by Max Kuhn and Julia Silge (O'Reilly). Copyright 2022 Max Kuhn and Julia Silge, 978-1-492-09648-1."

If you feel your use of code examples falls outside fair use or the permission given above, feel free to contact us at *permissions@oreilly.com*.

This version of the book was built with: R version 4.1.3 (2022-03-10), pandoc (*https://pandoc.org*) version 2.17.1.1, and the following packages:

- applicable (0.0.1.2, CRAN)
- av (0.7.0, CRAN)
- baguette (0.2.0, CRAN)
- beans (0.1.0, CRAN)
- bestNormalize (1.8.2, CRAN)
- bookdown (0.25, CRAN)
- broom (0.7.12, CRAN)
- censored (0.0.0.9000, GitHub)
- corrplot (0.92, CRAN)
- corrr (0.4.3, CRAN)
- Cubist (0.4.0, CRAN)
- DALEXtra (2.1.1, CRAN)
- dials (0.1.1, CRAN)
- dimRed (0.2.5, CRAN)
- discrim (0.2.0, CRAN)
- doMC (1.3.8, CRAN)
- dplyr (1.0.8, CRAN)
- earth (5.3.1, CRAN)
- embed (0.1.5, CRAN)
- fastICA (1.2-3, CRAN)
- finetune (0.2.0, CRAN)
- forcats (0.5.1, CRAN)
- ggforce (0.3.3, CRAN)
- ggplot2 (3.3.5, CRAN)
- glmnet (4.1-3, CRAN)
- gridExtra (2.3, CRAN)
- infer (1.0.0, CRAN)
- kableExtra (1.3.4, CRAN)
- kernlab (0.9-30, CRAN)
- kknn (1.3.1, CRAN)
- klaR (1.7-0, CRAN)
- knitr (1.38, CRAN)
- learntidymodels (0.0.0.9001, GitHub)
- lime (0.5.2, CRAN)
- lme4 (1.1-29, CRAN)
- lubridate (1.8.0, CRAN)
- mda (0.5-2, CRAN)
- mixOmics (6.18.1, Bioconductor)
- modeldata (0.1.1, CRAN)
- multilevelmod (0.1.0, CRAN)
- nlme (3.1-157, CRAN)
- nnet (7.3-17, CRAN)
- parsnip (0.2.1.9001, GitHub)
- patchwork (1.1.1, CRAN)
- pillar (1.7.0, CRAN)
- poissonreg (0.2.0, CRAN)
- prettyunits (1.1.1, CRAN)
- probably (0.0.6, CRAN)
- pscl (1.5.5, CRAN)
- purrr (0.3.4, CRAN)
- ranger (0.13.1, CRAN)
- recipes (0.2.0, CRAN)
- rlang (1.0.2, CRAN)
- rmarkdown (2.13, CRAN)
- rpart (4.1.16, CRAN)
- rsample (0.1.1, CRAN)
- rstanarm (2.21.3, CRAN)
- rules (0.2.0, CRAN)
- sessioninfo (1.2.2, CRAN)
- stacks (0.2.2, CRAN)

- stringr (1.4.0, CRAN)
- svglite (2.1.0, CRAN)
- text2vec (0.6, CRAN)
- textrecipes (0.5.1.9000, GitHub)
- themis (0.2.0, CRAN)
- tibble (3.1.6, CRAN)
- tidymodels (0.2.0, CRAN)
- tidyposterior (0.1.0, CRAN)

- tidyverse (1.3.1, CRAN)
- tune (0.2.0, CRAN)
- uwot (0.1.11, CRAN)
- workflows (0.2.6, CRAN)
- workflowsets (0.2.1, CRAN)
- xgboost (1.5.2.1, CRAN)
- yardstick (0.0.9, CRAN)

O'Reilly Online Learning

 For more than 40 years, *O'Reilly Media* has provided technology and business training, knowledge, and insight to help companies succeed.

Our unique network of experts and innovators share their knowledge and expertise through books, articles, and our online learning platform. O'Reilly's online learning platform gives you on-demand access to live training courses, in-depth learning paths, interactive coding environments, and a vast collection of text and video from O'Reilly and 200+ other publishers. For more information, visit *https://oreilly.com*.

How to Contact Us

Please address comments and questions concerning this book to the publisher:

O'Reilly Media, Inc.
1005 Gravenstein Highway North
Sebastopol, CA 95472
800-998-9938 (in the United States or Canada)
707-829-0515 (international or local)
707-829-0104 (fax)

We have a web page for this book, where we list errata, examples, and any additional information. You can access this page at *https://oreil.ly/tidy-modeling-r*.

Email *bookquestions@oreilly.com* to comment or ask technical questions about this book.

For news and information about our books and courses, visit *https://oreilly.com*.

Find us on LinkedIn: *https://linkedin.com/company/oreilly-media*.

Follow us on Twitter: *https://twitter.com/oreillymedia*.

Watch us on YouTube: *https://youtube.com/oreillymedia*.

Acknowledgments

We are so thankful for the contributions, help, and perspectives of people who have supported us in this project. There are several we would like to thank in particular.

We would like to thank our RStudio colleagues on the tidymodels team (Davis Vaughan, Hannah Frick, Emil Hvitfeldt, and Simon Couch) as well as the rest of our coworkers on the RStudio open source team. Thank you to Desirée De Leon for the site design of the online work. We would also like to thank our technical reviewers, Chelsea Parlett-Pelleriti and Dan Simpson, for their detailed, insightful feedback that substantively improved this book, as well as our editors, Nicole Taché and Rita Fernando, for their perspective and guidance during the process of writing and publishing.

This book was written in the open, and multiple people contributed via pull requests or issues. Special thanks goes to the 38 people who contributed via GitHub pull requests (in alphabetical order by username): Aris Paschalidis (@arisp99), Brad Hill (@bradisbrad), Bryce Roney (@bryceroney), Cedric Batailler (@cedricbatailler), Ildikó Czeller (@czeildi), David Kane (@davidkane9), @DavZim, @DCharIAA, Emil Hvitfeldt (@EmilHvitfeldt), Emilio (@emilopezcano), Fgazzelloni (@Fgazzelloni), Hannah Frick (@hfrick), Hlynur (@hlynurhallgrims), Howard Baek (@howardbaek), Jae Yeon Kim (@jaeyk), Jonathan D. Trattner (@jdtrat), Jeffrey Girard (@jmgirard), John W. Pickering (@JohnPickering), Jon Harmon (@jonthegeek), Joseph B. Rickert (@joseph-rickert), Maximilian Rohde (@maxdrohde), Michael Grund (@michaelgrund), @MikeJohnPage, Mine Cetinkaya-Rundel (@mine-cetinkaya-rundel), Mohammed Hamdy (@mmhamdy), @nattalides, Y. Yu (@PursuitOfDataScience), Riaz Hedayati (@riazhedayati), Rob Wiederstein (@RobWiederstein), Scott (@scottyd22), Simon Schölzel (@simonschoe), Simon Sayz (@tagasimon), @thrkng, Tanner Stauss (@tmstauss), Tony ElHabr (@tonyelhabr), Dmitry Zotikov (@x1o), Xiaochi (@xiaochi-liu), and Zach Bogart (@zachbogart).

## PART I

Introduction

## CHAPTER 1

Software for Modeling

Models are mathematical tools that can describe a system and capture relationships in the data given to them. Models can be used for various purposes, including predicting future events, determining if there is a difference between several groups, aiding map-based visualization, discovering novel patterns in the data that could be further investigated, and more. The utility of a model hinges on its ability to be reductive, or to reduce complex relationships to simpler terms. The primary influences in the data can be captured mathematically in a useful way, such as in a relationship that can be expressed as an equation.

Since the beginning of the 21st century, mathematical models have become ubiquitous in our daily lives, in both obvious and subtle ways. A typical day for many people might involve checking the weather to see when might be a good time to walk the dog, ordering a product from a website, typing a text message to a friend and having it autocorrected, and checking email. In each of these instances, there is a good chance that some type of model was involved. In some cases, the contribution of the model might be easily perceived ("You might also be interested in purchasing product X") while in other cases, the impact could be the absence of something (e.g., spam email). Models are used to choose clothing that a customer might like, to identify a molecule that should be evaluated as a drug candidate, and might even be the mechanism that a nefarious company uses to avoid the discovery of cars that overpollute. For better or worse, models are here to stay.

 There are two reasons that models permeate our lives today:

- An abundance of software exists to create models
- It has become easier to capture and store data, as well as make it accessible

This book focuses largely on software. It is obviously critical that software produces the correct relationships to represent the data. For the most part, determining mathematical correctness is possible, but the reliable creation of appropriate models requires more. In this chapter, we outline considerations for building or choosing modeling software, the purposes of models, and where modeling sits in the broader data analysis process.

Fundamentals for Modeling Software

It is important that the modeling software you use is easy to operate properly. The user interface should not be so poorly designed that the user would not know that they used it inappropriately. For example, Baggerly and Coombes (2009) report myriad problems in the data analyses from a high-profile computational biology publication. One of the issues was related to how the users were required to add the names of the model inputs. The software user interface made it easy to offset the column names of the data from the actual data columns. This resulted in the wrong genes being identified as important for treating cancer patients and eventually contributed to the termination of several clinical trials (Carlson 2012).

If we need high-quality models, software must facilitate proper usage. Abrams (2003) describes an interesting principle to guide us:

> The Pit of Success: in stark contrast to a summit, a peak, or a journey across a desert to find victory through many trials and surprises, we want our customers to simply fall into winning practices by using our platform and frameworks.

Data analysis and modeling software should espouse this idea.

Second, modeling software should promote good scientific methodology. When working with complex predictive models, it can be easy to unknowingly commit errors related to logical fallacies or inappropriate assumptions. Many machine learning models are so adept at discovering patterns that they can effortlessly find empirical patterns in the data that fail to reproduce later. Some of these methodological errors are insidious in that the issue can go undetected until a later time when new data that contain the true result are obtained.

As our models have become more powerful and complex, it has also become easier to commit latent errors.

This same principle also applies to programming. Whenever possible, the software should be able to protect users from committing mistakes. Software should make it easy for users to do the right thing.

These two aspects of model development—ease of proper use and good methodological practice—are crucial. Since tools for creating models are easily accessible and models can have such a profound impact, many more people are creating them. In terms of technical expertise and training, creators' backgrounds will vary. It is important that their tools be robust to the user's experience. Tools should be powerful enough to create high-performance models, but, on the other hand, should be easy to use appropriately. This book describes a suite of software for modeling that has been designed with these characteristics in mind.

The software is based on the R programming language (R Core Team 2014). R has been designed especially for data analysis and modeling. It is an implementation of the S language (with lexical scoping rules adapted from Scheme and Lisp) which was created in the 1970s to "turn ideas into software, quickly and faithfully" (Chambers 1998). R is open source and free. It is a powerful programming language that can be used for many different purposes but specializes in data analysis, modeling, visualization, and machine learning. R is easily extensible; it has a vast ecosystem of packages, mostly user-contributed modules that focus on a specific theme, such as modeling, visualization, and so on.

One collection of packages is called the *tidyverse* (Wickham et al. 2019). The tidyverse is an opinionated collection of R packages designed for data science. All packages share an underlying design philosophy, grammar, and data structures. Several of these design philosophies are directly informed by the aspects of software for modeling described in this chapter. If you've never used the tidyverse packages, Chapter 2 contains a review of basic concepts. Within the tidyverse, the subset of packages specifically focused on modeling are referred to as the *tidymodels* packages. This book is a practical guide for conducting modeling using the tidyverse and tidymodels packages. It shows how to use a set of packages, each with its own specific purpose, together to create high-quality models.

Types of Models

Before proceeding, let's describe a taxonomy for types of models, grouped by purpose. This taxonomy informs both how a model is used and many aspects of how the model may be created or evaluated. While this list is not exhaustive, most models fall into at least one of these categories: descriptive, inferential, or predictive.

Descriptive Models

The purpose of a descriptive model is to describe or illustrate characteristics of some data. The analysis might have no other purpose than to visually emphasize some trend or artifact in the data.

For example, large-scale measurements of RNA have been possible for some time using microarrays. Early laboratory methods placed a biological sample on a small microchip. Very small locations on the chip can measure a signal based on the abundance of a specific RNA sequence. The chip would contain thousands of outcomes (or more), each a quantification of the RNA related to a biological process. However, there could be quality issues on the chip that might lead to poor results. For example, a fingerprint accidentally left on a portion of the chip could cause inaccurate measurements when scanned.

An early method for evaluating such issues were probe-level models, or PLMs (Bolstad 2004). A statistical model would be created that accounted for the known differences in the data, such as the chip, the RNA sequence, the type of sequence, and so on. If there were other, unknown factors in the data, these effects would be captured in the model residuals. When the residuals were plotted by their location on the chip, a good quality chip would show no patterns. When a problem did occur, some sort of spatial pattern would be discernible. Often the type of pattern would suggest the underlying issue (e.g., a fingerprint) and a possible solution (wipe off the chip and rescan, repeat the sample, etc.). Figure 1-1(a) shows an application of this method for two microarrays taken from Gentleman et al. (2005). The images show two different color values; areas that are darker are where the signal intensity was larger than the model expects while the lighter color shows lower than expected values. The left-hand panel demonstrates a fairly random pattern while the right-hand panel exhibits an undesirable artifact in the middle of the chip.

Another example of a descriptive model is the *locally estimated scatterplot smoothing* model, more commonly known as LOESS (Cleveland 1979). Here, a smooth and flexible regression model is fit to a data set, usually with a single independent variable, and the fitted regression line is used to elucidate some trend in the data. These types of smoothers are used to discover potential ways to represent a variable in a model. This is demonstrated in Figure 1-1(b) where a nonlinear trend is illuminated by the flexible smoother. From this plot, it is clear that there is a highly nonlinear relationship between the sale price of a house and its latitude.

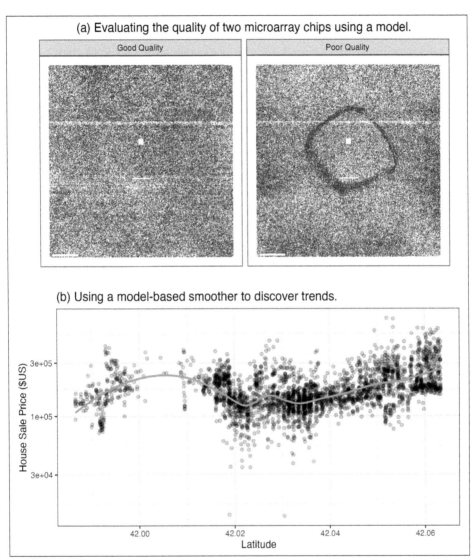

Figure 1-1. Two examples of how descriptive models can be used to illustrate specific patterns.

Inferential Models

The goal of an inferential model is to produce a decision for a research question or to explore a specific hypothesis, similar to how statistical tests are used.[1] An inferential model starts with a predefined conjecture or idea about a population and produces a statistical conclusion such as an interval estimate or the rejection of a hypothesis.

For example, the goal of a clinical trial might be to provide confirmation that a new therapy does a better job in prolonging life than an alternative, such as an existing therapy or no treatment at all. If the clinical endpoint is related to survival of a patient, the *null hypothesis* might be that the new treatment has an equal or lower median survival time, with the *alternative hypothesis* being that the new therapy has higher median survival. If this trial were evaluated using traditional null hypothesis significance testing via modeling, the significance testing would produce a p-value using some predefined methodology based on a set of assumptions for the data. Small values for the p-values in the model results would indicate there is evidence that the new therapy helps patients live longer. Large values for the p-values in the model results would conclude there is a failure to show such a difference; this lack of evidence could be due to a number of reasons, including the therapy not working.

What are the important aspects of this type of analysis? Inferential modeling techniques typically produce some type of probabilistic output, such as a p-value, confidence interval, or posterior probability. Generally, to compute such a quantity, formal probabilistic assumptions must be made about the data and the underlying processes that generated the data. The quality of the statistical modeling results are highly dependent on these predefined assumptions as well as how much the observed data appear to agree with them. The most critical factors here are theoretical: "If my data were independent and the residuals follow distribution X, then test statistic Y can be used to produce a p-value. Otherwise, the resulting p-value might be inaccurate."

 One aspect of inferential analyses is that there tends to be a delayed feedback loop in understanding how well the data match the model assumptions. In our clinical trial example, if statistical (and clinical) significance indicate that the new therapy should be available for patients to use, it still may be years before it is used in the field and enough data are generated for an independent assessment of whether the original statistical analysis led to the appropriate decision.

---

1 Many specific statistical tests are in fact equivalent to models. For example, t-tests and analysis of variance (ANOVA) methods are particular cases of the generalized linear model.

Predictive Models

Sometimes data are modeled to produce the most accurate prediction possible for new data. Here, the primary goal is that the predicted values have the highest possible fidelity to the true value of the new data.

A simple example would be for a book buyer to predict how many copies of a particular book should be shipped to their store for the next month. An overprediction wastes space and money due to excess books. If the prediction is smaller than it should be, there is opportunity loss and less profit.

For this type of model, the problem type is one of estimation rather than inference. For example, the buyer is usually not concerned with a question such as "Will I sell more than 100 copies of book X next month?" but rather "How many copies of book X will customers purchase next month?" Also, depending on the context, there may not be any interest in why the predicted value is X. In other words, there is more interest in the value itself than in evaluating a formal hypothesis related to the data. The prediction can also include measures of uncertainty. In the case of the book buyer, providing a forecasting error may be helpful in deciding how many books to purchase. It can also serve as a metric to gauge how well the prediction method worked.

What are the most important factors affecting predictive models? There are many different ways that a predictive model can be created, so the important factors depend on how the model was developed.[2]

A *mechanistic model* could be derived using first principles to produce a model equation that depends on assumptions. For example, when predicting the amount of a drug that's in a person's body at a certain time, some formal assumptions are made on how the drug is administered, absorbed, metabolized, and eliminated. Based on this, a set of differential equations is used to derive a specific model equation. Data are used to estimate the unknown parameters of this equation so that predictions can be generated. Like inferential models, mechanistic predictive models greatly depend on the assumptions that define their model equations. However, unlike inferential models, it's easy to make data-driven statements about how well the model performs based on how well it predicts the existing data. Here the feedback loop for the modeling practitioner is much faster than it would be for a hypothesis test.

Empirically driven models are created with more vague assumptions. These models tend to fall into the machine learning category. A good example is the K-nearest neighbor (KNN) model. Given a set of reference data, a new sample is predicted by using the values of the K most similar data in the reference set. For example,

---

2 Broader discussions of these distinctions can be found in Breiman (2001a) and Shmueli (2010).

if a book buyer needs a prediction for a new book, historical data from existing books may be available. A 5-nearest neighbor model would estimate the number of the new books to purchase based on the sales numbers of the five books that are most similar to the new one (for some definition of "similar"). This model is defined only by the structure of the prediction (the average of five similar books). No theoretical or probabilistic assumptions are made about the sales numbers or the variables that are used to define similarity. In fact, the primary method of evaluating the appropriateness of the model is to assess its accuracy using existing data. If the structure of this type of model were a good choice, the predictions would be close to the actual values.

Connections Between Types of Models

Note that we have defined the type of a model by how it is used, rather than its mathematical qualities.

An ordinary linear regression model might fall into any of these three classes of model, depending on how it is used:

- A descriptive smoother, similar to LOESS, called *restricted smoothing splines* (Durrleman and Simon 1989) can be used to describe trends in data using ordinary linear regression with specialized terms.

- An *analysis of variance* (ANOVA) model is a popular method for producing the p-values used for inference. ANOVA models are a special case of linear regression.

- If a simple linear regression model produces accurate predictions, it can be used as a predictive model.

There are many examples of predictive models that cannot (or at least should not) be used for inference. Even if probabilistic assumptions were made for the data, the nature of the K-nearest neighbors model, for example, makes the math required for inference intractable.

There is an additional connection between the types of models. While the primary purpose of descriptive and inferential models might not be related to prediction, the predictive capacity of the model should not be ignored. For example, logistic regression is a popular model for data in which the outcome is qualitative with two possible values. It can model how variables are related to the probability of the outcomes. When used inferentially, an abundance of attention is paid to the statistical qualities of the model. For example, analysts tend to strongly focus on the selection of independent variables contained in the model. Many iterations of model building may be used to determine a minimal subset of independent variables that have a *statistically significant* relationship to the outcome variable. This is usually achieved when all of the p-values for the independent variables are below a certain value (e.g., 0.05). From here, the analyst may focus on making qualitative statements about the relative influence that the variables have on the outcome (e.g., "There is a statistically significant relationship between age and the odds of heart disease.").

However, this approach can be dangerous when statistical significance is used as the only measure of model quality. It is possible that this statistically optimized model has poor model accuracy, or it performs poorly on some other measure of predictive capacity. While the model might not be used for prediction, how much should inferences be trusted from a model that has significant p-values but dismal accuracy? Predictive performance tends to be related to how close the model's fitted values are to the observed data.

 If a model has limited fidelity to the data, the inferences generated by the model should be highly suspect. In other words, statistical significance may not be sufficient proof that a model is appropriate.

This may seem intuitively obvious, but it is often ignored in real-world data analysis.

Some Terminology

Before proceeding, we will outline additional terminology related to modeling and data. These descriptions are intended to be helpful as you read this book, but they are not exhaustive.

First, many models can be categorized as being *supervised* or *unsupervised*. Unsupervised models are those that learn patterns, clusters, or other characteristics of the data but lack an outcome, i.e., a dependent variable. Principal component analysis (PCA), clustering, and autoencoders are examples of unsupervised models; they are used to understand relationships between variables or sets of variables without an explicit relationship between predictors and an outcome. Supervised models are those that have an outcome variable. Linear regression, neural networks, and numerous other methodologies fall into this category.

Within supervised models, there are two main subcategories:

- *Regression* predicts a numeric outcome.
- *Classification* predicts an outcome that is an ordered or unordered set of qualitative values.

These are imperfect definitions and do not account for all possible model types. In Chapter 6, we refer to this characteristic of supervised techniques as the *model mode*.

Different variables can have different *roles*, especially in a supervised modeling analysis. Outcomes (otherwise known as the labels, endpoints, or dependent variables) are the value being predicted in supervised models. The independent variables, which are the substrate for making predictions of the outcome, are also referred to as predictors, features, or covariates (depending on the context). The terms *outcomes* and *predictors* are used most frequently in this book.

In terms of the data or variables themselves, whether used for supervised or unsupervised models, as predictors or outcomes, the two main categories are quantitative and qualitative. Examples of the former are real numbers like 3.14159 and integers like 42. Qualitative values, also known as nominal data, are those that represent some sort of discrete state that cannot be naturally placed on a numeric scale, like "red," "green," and "blue."

How Does Modeling Fit into the Data Analysis Process?

In what circumstances are models created? Are there steps that precede such an undertaking? Is model creation the first step in data analysis?

There are a few critical phases of data analysis that always come before modeling.

First, there is the chronically underestimated process of *cleaning the data*. No matter the circumstances, you should investigate the data to make sure that they are applicable to your project goals, accurate, and appropriate. These steps can easily take more time than the rest of the data analysis process (depending on the circumstances).

Data cleaning can also overlap with the second phase of *understanding the data*, often referred to as exploratory data analysis (EDA). EDA brings to light how the different variables are related to one another, their distributions, typical ranges, and other attributes. A good question to ask at this phase is "How did I come by *these* data?" This question can help you understand how the data at hand have been sampled or filtered and if these operations were appropriate. For example, when merging database tables, a join may go awry, which could accidentally eliminate one or more subpopulations. Another good idea is to ask if the data are relevant. For example, to predict whether patients have Alzheimer's disease, it would be unwise to have a data set containing subjects with the disease and a random sample of healthy adults from the general population. Given the progressive nature of the disease, the model may simply predict who are the oldest patients.

Finally, before starting a data analysis process, there should be clear expectations of the model's goal and how performance (and success) will be judged. At least one *performance metric* should be identified with realistic goals of what can be achieved. Common statistical metrics, discussed in more detail in Chapter 9, are classification accuracy, true and false positive rates, root mean squared error, and so on. The relative benefits and drawbacks of these metrics should be weighed. It is also important that the metric be germane; alignment with the broader data analysis goals is critical.

The process of investigating the data may not be simple. Wickham and Grolemund (2016) contains an excellent illustration of the general data analysis process, reproduced in Figure 1-2. Data ingestion and cleaning/tidying are shown as the initial steps. When the analytical steps for understanding commence, they are a heuristic process; we cannot predetermine how long they may take. The cycle of transformation, modeling, and visualization often requires multiple iterations.

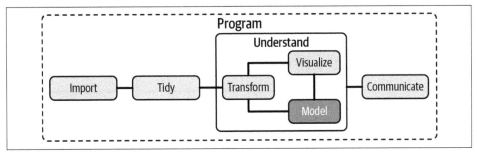

Figure 1-2. The data science process.

This iterative process is especially true for modeling. Figure 1-3 emulates the typical path to determining an appropriate model. The general phases are:

Exploratory data analysis (EDA)
Initially there is a back and forth between numerical analysis and data visualization (represented in Figure 1-2) in which different discoveries lead to more questions and data analysis side-quests to gain more understanding.

Feature engineering
The understanding gained from EDA results in the creation of specific model terms that make it easier to accurately model the observed data. This can include complex methodologies (e.g., PCA) or simpler features (using the ratio of two predictors). Chapter 8 focuses entirely on this important step.

Model tuning and selection (large circles with alternating segments)

A variety of models are generated and their performance is compared. Some models require parameter tuning in which some structural parameters must be specified or optimized. The alternating segments within the circles signify the repeated data splitting used during resampling (see Chapter 10).

Model evaluation

During this phase of model development, we assess the model's performance metrics, examine residual plots, and conduct other EDA-like analyses to understand how well the models work. In some cases, formal between-model comparisons (Chapter 11) help you understand whether any differences in models are within the experimental noise.

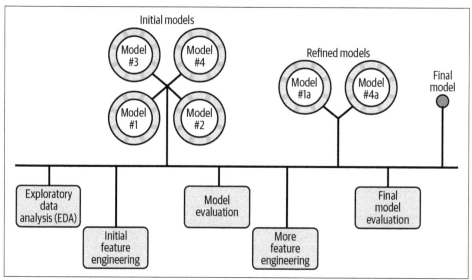

Figure 1-3. A schematic for the typical modeling process.

After an initial sequence of these tasks, more understanding is gained regarding which models are superior as well as which data subpopulations are not being effectively estimated. This leads to additional EDA and feature engineering, another round of modeling, and so on. Once the data analysis goals are achieved, typically the last steps are to finalize, document, and communicate the model. For predictive models, it is common at the end to validate the model on an additional set of data reserved for this specific purpose.

As an example, Kuhn and Johnson (2020) use data to model the daily ridership of Chicago's public train system using predictors such as the date, the previous ridership results, the weather, and other factors. Table 1-1 shows an approximation of these authors' hypothetical inner monologue when analyzing these data and eventually selecting a model with sufficient performance.

Table 1-1. Hypothetical inner monologue of a model developer

Thoughts	Activity
The daily ridership values between stations are extremely correlated.	EDA
Weekday and weekend ridership look very different.	EDA
One day in the summer of 2010 has an abnormally large number of riders.	EDA
Which stations had the lowest daily ridership values?	EDA
Dates should at least be encoded as day-of-the-week, and year.	Feature engineering
Maybe PCA could be used on the correlated predictors to make it easier for the models to use them.	Feature engineering
Hourly weather records should probably be summarized into daily measurements.	Feature engineering
Let's start with simple linear regression, *K*-nearest neighbors, and a boosted decision tree.	Model fitting
How many neighbors should be used?	Model tuning
Should we run a lot of boosting iterations or just a few?	Model tuning
How many neighbors seemed to be optimal for these data?	Model tuning
Which models have the lowest root mean squared errors?	Model evaluation
Which days were poorly predicted?	EDA
Variable importance scores indicate that the weather information is not predictive. We'll drop them from the next set of models.	Model evaluation
It seems like we should focus on a lot of boosting iterations for that model.	Model evaluation
We need to encode holiday features to improve predictions on (and around) those dates.	Feature engineering
Let's drop KNN from the model list.	Model evaluation

Chapter Summary

This chapter focused on how models describe relationships in data, as well as different types of models such as descriptive models, inferential models, and predictive models. The predictive capacity of a model can be used to evaluate it, even when its main goal is not prediction. Modeling itself sits within the broader data analysis process, and exploratory data analysis is a key part of building high-quality models.

## CHAPTER 2

A Tidyverse Primer

What is the tidyverse, and where does the tidymodels framework fit in? The tidyverse is a collection of R packages for data analysis that are developed with common ideas and norms. From Wickham et al. (2019):

> At a high level, the tidyverse is a language for solving data science challenges with R code. Its primary goal is to facilitate a conversation between a human and a computer about data. Less abstractly, the tidyverse is a collection of R packages that share a high-level design philosophy and low-level grammar and data structures, so that learning one package makes it easier to learn the next.

In this chapter, we briefly discuss important principles of the tidyverse design philosophy and how they apply in the context of modeling software that is easy to use properly and supports good statistical practice, like we outlined in Chapter 1. The next chapter covers modeling conventions from the core R language. Together, you can use these discussions to understand the relationships between the tidyverse, tidymodels, and the core or base R language. Both tidymodels and the tidyverse build on the R language, and tidymodels applies tidyverse principles to building models.

Tidyverse Principles

The full set of strategies and tactics for writing R code in the tidyverse style can be found at the tidyverse website (*https://oreil.ly/09DEV*). Here we briefly describe several of the general tidyverse design principles, their motivation, and how we think about modeling as an application of these principles.

Design for Humans

The tidyverse focuses on designing R packages and functions that can be easily understood and used by a broad range of people. Both historically and today, a substantial percentage of R users are not people who create software or tools but instead people who create analyses or models. As such, R users do not typically have (or need) computer science backgrounds, and many are not interested in writing their own R packages.

For this reason, it is critical that R code be easy to work with to accomplish your goals. Documentation, training, accessibility, and other factors play an important part in achieving this. However, if the syntax itself is difficult for people to easily comprehend, documentation is a poor solution. The software itself must be intuitive.

To contrast the tidyverse approach with more traditional R semantics, consider sorting a data frame. Data frames can represent different types of data in each column, and multiple values in each row. Using only the core language, we can sort a data frame using one or more columns by reordering the rows via R's subscripting rules in conjunction with `order()`; you cannot successfully use a function you might be tempted to try in such a situation because of its name, `sort()`. To sort the `mtcars` data by two of its columns, the call might look like:

```
mtcars[order(mtcars$gear, mtcars$mpg), ]
```

While very computationally efficient, it would be difficult to argue that this is an intuitive user interface. In dplyr by contrast, the tidyverse function `arrange()` takes a set of variable names as input arguments directly:

```
library(dplyr)
arrange(.data = mtcars, gear, mpg)
```

 The variable names used here are "unquoted"; many traditional R functions require a character string to specify variables, but tidyverse functions take unquoted names or *selector functions*. The selectors allow for one or more readable rules that are applied to the column names. For example, `ends_with("t")` would select the `drat` and `wt` columns of the `mtcars` data frame.

Additionally, naming is crucial. If you were new to R and were writing data analysis or modeling code involving linear algebra, you might be stymied when searching for a function that computes the matrix inverse. Using `apropos("inv")` yields no candidates. It turns out that the base R function for this task is `solve()`, for solving systems of linear equations. For a matrix X, you would use `solve(X)` to invert X (with no vector for the right-hand side of the equation). This is only documented in the description of one of the *arguments* in the help file. In essence, you need to know the name of the solution to be able to find the solution.

The tidyverse approach is to use function names that are descriptive and explicit over those that are short and implicit. There is a focus on verbs (e.g., fit, arrange, etc.) for general methods. Verb-noun pairs are particularly effective; consider invert_matrix() as a hypothetical function name. In the context of modeling, it is also important to avoid highly technical jargon, such as Greek letters or obscure terms, in names. Names should be as self-documenting as possible.

When there are similar functions in a package, function names are designed to be optimized for tab-completion. For example, the glue package has a collection of functions starting with a common prefix (glue_) that enables users to quickly find the function they are looking for.

Reuse Existing Data Structures

Whenever possible, functions should avoid returning a novel data structure. If the results are conducive to an existing data structure, it should be used. This reduces the cognitive load when using software; no additional syntax or methods are required.

The data frame is the preferred data structure in tidyverse and tidymodels packages because its structure is a good fit for such a broad swath of data science tasks. Specifically, the tidyverse and tidymodels favor the *tibble*, a modern reimagining of R's data frame that we describe in the next section on example tidyverse syntax.

As an example, the rsample package can be used to create *resamples* of a data set, such as cross-validation or the bootstrap (described in Chapter 10). The resampling functions return a tibble with a column (called splits) of objects that define the resampled data sets. Three bootstrap samples of a data set might look like:

```
boot_samp <- rsample::bootstraps(mtcars, times = 3)
boot_samp
#> # Bootstrap sampling
#> # A tibble: 3 × 2
#>   splits          id
#>   <list>          <chr>
#> 1 <split [32/15]> Bootstrap1
#> 2 <split [32/12]> Bootstrap2
#> 3 <split [32/10]> Bootstrap3
class(boot_samp)
#> [1] "bootstraps" "rset"       "tbl_df"     "tbl"        "data.frame"
```

With this approach, vector-based functions can be used with these columns, such as vapply() or purrr::map().[1] This boot_samp object has multiple classes but inherits methods for data frames ("data.frame") and tibbles ("tbl_df"). Additionally, new columns can be added to the results without affecting the class of the data. This is

---

1 If you've never seen :: in R code, it's an explicit method for calling a function. The value of the left-hand side is the *namespace* where the function lives (usually a package name). The right-hand side is the function name. When two packages use the same function name, this syntax ensures that the correct function is called.

much easier and more versatile for users to work with than a completely new object type that does not make its data structure obvious.

One downside to relying on common data structures is the potential loss of computational performance. In some situations, data can be encoded in specialized formats that are more efficient representations of the data. For example:

- In computational chemistry, the structure-data file format (SDF) is a tool to take chemical structures and encode them in a format that is computationally efficient to work with.

- Data that have a large number of values that are the same (such as zeros for binary data) can be stored in a sparse matrix format. This format can reduce the size of the data as well as enable more efficient computational techniques.

These formats are advantageous when the problem is well scoped and the potential data processing methods are both well defined and suited to such a format.[2] However, once such constraints are violated, specialized data formats are less useful. For example, if we perform a transformation of the data that converts the data into fractional numbers, the output is no longer sparse; the sparse matrix representation is helpful for one specific algorithmic step in modeling, but this is often not true before or after that specific step.

 A specialized data structure is not flexible enough for an entire modeling workflow in the way that a common data structure is.

One important feature in the tibble produced by rsample is that the `splits` column is a list. In this instance, each element of the list has the same type of object: an `rsplit` object that contains the information about which rows of `mtcars` belong in the bootstrap sample. *List columns* can be very useful in data analysis and, as will be seen throughout this book, are important to tidymodels.

Design for the Pipe and Functional Programming

The magrittr pipe operator (`%>%`) is a tool for chaining together a sequence of R functions.[3] To demonstrate, consider the following commands that sort a data frame and then retain the first 10 rows:

---

2 Not all algorithms can take advantage of sparse representations of data. In such cases, a sparse matrix must be converted to a more conventional format before proceeding.

```
small_mtcars <- arrange(mtcars, gear)
small_mtcars <- slice(small_mtcars, 1:10)

# or more compactly:
small_mtcars <- slice(arrange(mtcars, gear), 1:10)
```

The pipe operator substitutes the value of the left-hand side of the operator as the first argument to the right-hand side, so we can implement the same result as before with:

```
small_mtcars <-
  mtcars %>%
  arrange(gear) %>%
  slice(1:10)
```

The piped version of this sequence is more readable; this readability increases as more operations are added to a sequence. This approach to programming works in this example because all of the functions we used return the same data structure (a data frame) that is then the first argument to the next function. This is by design. When possible, create functions that can be incorporated into a pipeline of operations.

If you have used ggplot2, this is not unlike the layering of plot components into a ggplot object with the + operator. To make a scatterplot with a regression line, the initial ggplot() call is augmented with two additional operations:

```
library(ggplot2)
ggplot(mtcars, aes(x = wt, y = mpg)) +
  geom_point() +
  geom_smooth(method = lm)
```

While similar to the dplyr pipeline, note that the first argument to this pipeline is a data set (mtcars) and that each function call returns a ggplot object. Not all pipelines need to keep the returned values (plot objects) the same as the initial value (a data frame). Using the pipe operator with dplyr operations has acclimated many R users to expect to return a data frame when pipelines are used; as shown with ggplot2, this does not need to be the case. Pipelines are incredibly useful in modeling workflows, but modeling pipelines can return, instead of a data frame, objects such as model components.

R has excellent tools for creating, changing, and operating on functions, making it a great language for functional programming. This approach can replace iterative loops in many situations, such as when a function returns a value without other side effects.[4]

---

3 In R 4.1, a native pipe operator |> was introduced. In this book, we use the magrittr pipe since users of older versions of R will not have the new native pipe.

4 Examples of function side effects could include changing global data or printing a value.

Let's look at an example. Suppose you are interested in the logarithm of the ratio of the fuel efficiency to the car weight. To those new to R and/or coming from other programming languages, a loop might seem like a good option:

```
n <- nrow(mtcars)
ratios <- rep(NA_real_, n)
for (car in 1:n) {
  ratios[car] <- log(mtcars$mpg[car]/mtcars$wt[car])
}
head(ratios)
#> [1] 2.081 1.988 2.285 1.896 1.693 1.655
```

Those with more experience in R may know that there is a much simpler and faster vectorized version that can be computed by:

```
ratios <- log(mtcars$mpg/mtcars$wt)
```

However, in many real-world cases, the element-wise operation of interest is too complex for a vectorized solution. In such a case, a good approach is to write a function to do the computations. When we design for functional programming, it is important that the output depends only on the inputs and that the function has no side effects. Violations of these ideas in the following function are shown with comments:

```
compute_log_ratio <- function(mpg, wt) {
  log_base <- getOption("log_base", default = exp(1)) # gets external data
  results <- log(mpg/wt, base = log_base)
  print(mean(results))                              # prints to the console
  done <<- TRUE                                     # sets external data
  results
}
```

A better version would be:

```
compute_log_ratio <- function(mpg, wt, log_base = exp(1)) {
  log(mpg/wt, base = log_base)
}
```

The purrr package contains tools for functional programming. Let's focus on the map() family of functions, which operates on vectors and always returns the same type of output. The most basic function, map(), always returns a list and uses the basic syntax of map(vector, function). For example, to take the square root of our data, we could:

```
map(head(mtcars$mpg, 3), sqrt)
#> [[1]]
#> [1] 4.583
#>
#> [[2]]
#> [1] 4.583
#>
#> [[3]]
#> [1] 4.775
```

There are specialized variants of map() that return values when we know or expect that the function will generate one of the basic vector types. For example, since the square root returns a double-precision number:

```
map_dbl(head(mtcars$mpg, 3), sqrt)
#> [1] 4.583 4.583 4.775
```

There are also mapping functions that operate across multiple vectors:

```
log_ratios <- map2_dbl(mtcars$mpg, mtcars$wt, compute_log_ratio)
head(log_ratios)
#> [1] 2.081 1.988 2.285 1.896 1.693 1.655
```

The map() functions also allow for temporary, anonymous functions defined using the tilde character. The argument values are .x and .y for map2():

```
map2_dbl(mtcars$mpg, mtcars$wt, ~ log(.x/.y)) %>%
  head()
#> [1] 2.081 1.988 2.285 1.896 1.693 1.655
```

These examples have been trivial but, in later sections, will be applied to more complex problems.

 For functional programming in tidy modeling, functions should be defined so that functions like map() can be used for iterative computations.

Examples of Tidyverse Syntax

Let's begin our discussion of tidyverse syntax by exploring more deeply what a tibble is and how tibbles work. Tibbles have slightly different rules than basic data frames in R. For example, tibbles naturally work with column names that are not syntactically valid variable names:

```
# Wants valid names:
data.frame(`variable 1` = 1:2, two = 3:4)
#>   variable.1 two
#> 1          1   3
#> 2          2   4
# But can be coerced to use them with an extra option:
df <- data.frame(`variable 1` = 1:2, two = 3:4, check.names = FALSE)
df
#>   variable 1 two
#> 1          1   3
#> 2          2   4

# But tibbles just work:
tbbl <- tibble(`variable 1` = 1:2, two = 3:4)
tbbl
#> # A tibble: 2 × 2
#>   `variable 1`   two
#>          <int> <int>
```

```
#> 1          1    3
#> 2          2    4
```

Standard data frames enable *partial matching* of arguments so that code using only a portion of the column names still works. Tibbles prevent this from happening since it can lead to accidental errors:

```
df$tw
#> [1] 3 4

tbbl$tw
#> Warning: Unknown or uninitialized column: `tw`.
#> NULL
```

Tibbles also prevent one of the most common R errors: dropping dimensions. If a standard data frame subsets the columns down to a single column, the object is converted to a vector. Tibbles never do this:

```
df[, "two"]
#> [1] 3 4

tbbl[, "two"]
#> # A tibble: 2 × 1
#>     two
#>   <int>
#> 1     3
#> 2     4
```

There are other advantages to using tibbles instead of data frames, such as better printing and more.[5]

To demonstrate some syntax, let's use tidyverse functions to read in data that could be used in modeling. The data set comes from the city of Chicago's data portal and contains daily ridership data for the city's elevated train stations. The data set has columns for:

- The station identifier (numeric)
- The station name (character)
- The date (character in mm/dd/yyyy format)
- The day of the week (character)
- The number of riders (numeric)

Our tidyverse pipeline will conduct the following tasks, in order:

1. Use the tidyverse package readr to read the data from the source website and convert them into a tibble. To do this, the `read_csv()` function can determine the type of data by reading an initial number of rows. Alternatively, if the column

---

5 Chapter 10 of Wickham and Grolemund (2016) has more details on tibbles.

names and types are already known, a column specification can be created in R and passed to `read_csv()`.

2. Filter the data to eliminate a few columns that are not needed (such as the station ID) and change the column `stationname` to `station`. The function `select()` is used for this. When filtering, use either the column names or a dplyr selector function. When selecting names, a new variable name can be declared using the argument format `new_name = old_name`.

3. Convert the date field to the R data format using the `mdy()` function from the lubridate package. We also convert the ridership numbers to thousands. Both of these computations are executed using the `dplyr::mutate()` function.

4. Use the maximum number of rides for each station and day combination. This mitigates the issue of a small number of days that have more than one record of ridership numbers at certain stations. We group the ridership data by station and day, and then summarize within each of the 1,999 unique combinations with the maximum statistic.

The tidyverse code for these steps is:

```
library(tidyverse)
library(lubridate)

url <- "http://bit.ly/raw-train-data-csv"

all_stations <-
  # Step 1: Read in the data.
  read_csv(url) %>%
  # Step 2: filter columns and rename stationname
  dplyr::select(station = stationname, date, rides) %>%
  # Step 3: Convert the character date field to a date encoding.
  # Also, put the data in units of 1K rides
  mutate(date = mdy(date), rides = rides / 1000) %>%
  # Step 4: Summarize the multiple records using the maximum.
  group_by(date, station) %>%
  summarize(rides = max(rides), .groups = "drop")
```

This pipeline of operations illustrates why the tidyverse is popular. The pipeline uses a series of data manipulations that have simple and easy to understand functions for each transformation; the series is bundled in a streamlined, readable way. The focus is on how the user interacts with the software. This approach enables more people to learn R and achieve their analysis goals, and adopting these same principles for modeling in R has the same benefits.

Chapter Summary

This chapter introduced the tidyverse, with a focus on applications for modeling and how tidyverse design principles inform the tidymodels framework. Think of the tidymodels framework as applying tidyverse principles to the domain of building models. We described differences in conventions between the tidyverse and base R, and introduced two important components of the tidyverse system: tibbles and the pipe operator %>%. Data cleaning and processing can feel mundane at times, but these tasks are important for modeling in the real world; we illustrated how to use tibbles, the pipe, and tidyverse functions in an example data import and processing exercise.

## CHAPTER 3

A Review of R Modeling Fundamentals

Before describing how to use tidymodels for applying tidy data principles to building models with R, let's review how models are created, trained, and used in the core R language (often called base R). This chapter is a brief illustration of core language conventions that are important to be aware of even if you never use base R for models at all. This chapter is not exhaustive, but it provides readers (especially those new to R) the basic, most commonly used motifs.

The S language, on which R is based, has had a rich data analysis environment since the publication of Chambers and Hastie (1992) (commonly known as The White Book). This version of S introduced standard infrastructure components familiar to R users today, such as symbolic model formulas, model matrices, and data frames, as well as standard object-oriented programming methods for data analysis. These user interfaces have not substantively changed since then.

An Example

To demonstrate some fundamentals for modeling in base R, let's use experimental data from McDonald (2009), by way of Mangiafico (2015), on the relationship between the ambient temperature and the rate of cricket chirps per minute. Data were collected for two species: *O. exclamationis* and *O. niveus*. The data are contained in a data frame called `crickets` with a total of 31 data points. These data are shown in Figure 3-1 using the following ggplot2 code:

```
library(tidyverse)

data(crickets, package = "modeldata")
names(crickets)

# Plot the temperature on the x-axis, the chirp rate on the y-axis. The plot
# elements will be colored differently for each species:
ggplot(crickets,
```

```
        aes(x = temp, y = rate, color = species, pch = species, lty = species)) +
      # Plot points for each data point and color by species
      geom_point(size = 2) +
      # Show a simple linear model fit created separately for each species:
      geom_smooth(method = lm, se = FALSE, alpha = 0.5) +
      scale_color_brewer(palette = "Paired") +
      labs(x = "Temperature (C)", y = "Chirp Rate (per minute)")

   #> [1] "species" "temp"    "rate"
```

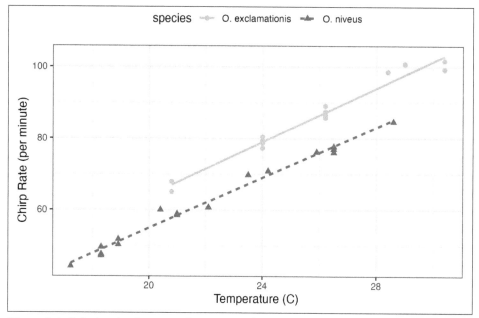

Figure 3-1. Relationship between chirp rate and temperature for two species of cricket.

The data exhibit fairly linear trends for each species. For a given temperature, *O. exclamationis* appears to chirp more per minute than the other species. For an inferential model, the researchers might have specified the following null hypotheses prior to seeing the data:

- Temperature has no effect on the chirp rate.
- There are no differences between the species' chirp rates.

There may be some scientific or practical value in predicting the chirp rate, but in this example we will focus on inference.

To fit an ordinary linear model in R, the lm() function is commonly used. The important arguments to this function are a model formula and a data frame that contains the data. The formula is *symbolic*. For example, the simple formula:

```
   rate ~ temp
```

specifies that the chirp rate is the outcome (since it's on the left-hand side of the tilde ~) and the temperature value is the predictor.[1]

Suppose the data contained the time of day in which the measurements were obtained in a column called time. The formula:

```
rate ~ temp + time
```

would not add the time and temperature values together. This formula would symbolically represent that temperature and time should be added as separate *main effects* to the model. A main effect is a model term that contains a single predictor variable.

There are no time measurements in these data, but the species can be added to the model in the same way:

```
rate ~ temp + species
```

Species is not a quantitative variable; in the data frame, it is represented as a factor column with levels "O. exclamationis" and "O. niveus". The vast majority of model functions cannot operate on nonnumeric data. For species, the model needs to encode the species data in a numeric format. The most common approach is to use *indicator variables* (also known as dummy variables) in place of the original qualitative values. In this instance, since species has two possible values, the model formula will automatically encode this column as numeric by adding a new column that has a value of zero when the species is "O. exclamationis" and a value of one when the data correspond to "O. niveus". The underlying formula machinery automatically converts these values for the data set used to create the model, as well as for any new data points (for example, when the model is used for prediction).

 Suppose there were five species instead of two. The model formula would automatically add four additional binary columns that are binary indicators for four of the species. The *reference level* of the factor (i.e., the first level) is always left out of the predictor set. The idea is that, if you know the values of the four indicator variables, the value of the species can be determined. We discuss binary indicator variables in more detail in Chapter 8.

The model formula rate ~ temp + species creates a model with different y-intercepts for each species; the slopes of the regression lines could be different for each species as well. To accommodate this structure, an interaction term can be added to the model. This can be specified in a few different ways, and the most basic uses the colon:

---

1 Most model functions implicitly add an intercept column.

```
rate ~ temp + species + temp:species

# A shortcut can be used to expand all interactions containing
# interactions with two variables:
rate ~ (temp + species)^2

# Another shortcut to expand factors to include all possible
# interactions (equivalent for this example):
rate ~ temp * species
```

In addition to the convenience of automatically creating indicator variables, the formula offers a few other niceties:

- *In-line* functions can be used in the formula. For example, to use the natural log of the temperature, we can create the formula `rate ~ log(temp)`. Since the formula is symbolic by default, literal math can also be applied to the predictors using the identity function `I()`. To use Fahrenheit units, the formula could be `rate ~ I( (temp * 9/5) + 32 )` to convert from Celsius.

- R has many functions that are useful inside of formulas. For example, `poly(x, 3)` creates linear, quadratic, and cubic terms for x to the model as main effects. The splines package also has several functions to create nonlinear spline terms in the formula.

- For data sets where there are many predictors, the period shortcut is available. The period represents main effects for all of the columns that are not on the left-hand side of the tilde. Using `~ (.)^3` would create main effects as well as all two- and three-variable interactions to the model.

Returning to our chirping crickets, let's use a two-way interaction model. In this book, we use the suffix `_fit` for R objects that are fitted models:

```
interaction_fit <-  lm(rate ~ (temp + species)^2, data = crickets)

# To print a short summary of the model:
interaction_fit
#>
#> Call:
#> lm(formula = rate ~ (temp + species)^2, data = crickets)
#>
#> Coefficients:
#>         (Intercept)                 temp      species0. niveus
#>             -11.041                3.751               -4.348
#> temp:species0. niveus
#>              -0.234
```

This output is a little hard to read. For the species indicator variables, R mashes the variable name (`species`) together with the factor level (`0. niveus`) with no delimiter.

Before going into any inferential results for this model, the fit should be assessed using diagnostic plots. We can use the `plot()` method for `lm` objects. This method

---

produces a set of four plots for the object, each showing different aspects of the fit, as shown in Figure 3-2:

```
# Place two plots next to one another:
par(mfrow = c(1, 2))

# Show residuals versus predicted values:
plot(interaction_fit, which = 1)

# A normal quantile plot on the residuals:
plot(interaction_fit, which = 2)
```

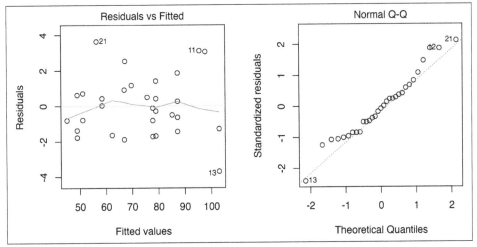

Figure 3-2. Residual diagnostic plots for the linear model with interactions, which appear reasonable enough to conduct inferential analysis.

When it comes to the technical details of evaluating expressions, R is *lazy* (as opposed to eager). This means that model fitting functions typically compute the minimum possible quantities at the last possible moment. For example, if you are interested in the coefficient table for each model term, this is not automatically computed with the model but is instead computed via the summary() method.

Our next order of business with the crickets is to assess if the inclusion of the interaction term is necessary. The most appropriate approach for this model is to recompute the model without the interaction term and use the anova() method:

```
# Fit a reduced model:
main_effect_fit <-  lm(rate ~ temp + species, data = crickets)

# Compare the two:
anova(main_effect_fit, interaction_fit)
#> Analysis of Variance Table
#>
#> Model 1: rate ~ temp + species
```

```
#> Model 2: rate ~ (temp + species)^2
#>   Res.Df  RSS Df Sum of Sq    F Pr(>F)
#> 1    28 89.3
#> 2    27 85.1  1    4.28 1.36   0.25
```

This statistical test generates a p-value of 0.25. This implies that there is a lack of evidence against the null hypothesis that the interaction term is not needed by the model. For this reason, we will conduct further analysis on the model without the interaction.

Residual plots should be reassessed to make sure that our theoretical assumptions are valid enough to trust the p-values produced by the model (plots are not shown here, but spoiler alert: they are).

We can use the `summary()` method to inspect the coefficients, standard errors, and p-values of each model term:

```
summary(main_effect_fit)
#>
#> Call:
#> lm(formula = rate ~ temp + species, data = crickets)
#>
#> Residuals:
#>     Min      1Q Median      3Q     Max
#> -3.013 -1.130 -0.391  0.965   3.780
#>
#> Coefficients:
#>                    Estimate Std. Error t value Pr(>|t|)
#> (Intercept)         -7.2109     2.5509   -2.83   0.0086 **
#> temp                 3.6028     0.0973   37.03  < 2e-16 ***
#> species0. niveus   -10.0653     0.7353  -13.69  6.3e-14 ***
#> ---
#> Signif. codes:  0 '***' 0.001 '**' 0.01 '*' 0.05 '.' 0.1 ' ' 1
#>
#> Residual standard error: 1.79 on 28 degrees of freedom
#> Multiple R-squared:  0.99,   Adjusted R-squared:  0.989
#> F-statistic: 1.33e+03 on 2 and 28 DF,  p-value: <2e-16
```

The chirp rate for each species increases by 3.6 chirps as the temperature increases by a single degree. This term shows strong statistical significance as evidenced by the p-value. The species term has a value of –10.07. This indicates that, across all temperature values, O. niveus has a chirp rate that is about 10 fewer chirps per minute than O. exclamationis. Similar to the temperature term, the species effect is associated with a very small p-value.

The only issue in this analysis is the intercept value. It indicates that at 0°C, there are negative chirps per minute for both species. While this doesn't make sense, the data only go as low as 17.2°C and interpreting the model at 0°C would be an extrapolation. This would be a bad idea. That being said, the model fit is good within the *applicable range* of the temperature values; the conclusions should be limited to the observed temperature range.

If we needed to estimate the chirp rate at a temperature that was not observed in the experiment, we could use the `predict()` method. It takes the model object and a data frame of new values for prediction. For example, the model estimates the chirp rate for *O. exclamationis* for temperatures between 15°C and 20°C can be computed via:

```
new_values <- data.frame(species = "O. exclamationis", temp = 15:20)
predict(main_effect_fit, new_values)
#>     1     2     3     4     5     6
#> 46.83 50.43 54.04 57.64 61.24 64.84
```

 Note that the nonnumeric value of `species` is passed to the predict method, as opposed to the numeric, binary indicator variable.

While this analysis has obviously not been an exhaustive demonstration of R's modeling capabilities, it does highlight some major features important for the rest of this book:

- The language has an expressive syntax for specifying model terms for both simple and quite complex models.
- The R formula method has many conveniences for modeling that are also applied to new data when predictions are generated.
- There are numerous helper functions (e.g., `anova()`, `summary()`, and `predict()`) that you can use to conduct specific calculations after the fitted model is created.

Finally, as previously mentioned, this framework was first published in 1992. Most of these ideas and methods were developed in that period but have remained remarkably relevant to this day. It highlights that the S language, and by extension R, has been designed for data analysis since its inception.

What Does the R Formula Do?

The R model formula is used by many modeling packages. It usually serves multiple purposes:

- The formula defines the columns that the model uses.
- The standard R machinery uses the formula to encode the columns into an appropriate format.
- The roles of the columns are defined by the formula.

For the most part, practitioners' understanding of what the formula does is dominated by the last purpose. Our focus when typing out a formula is often to declare how the columns should be used. For example, the previous specification we discussed sets up predictors to be used in a specific way:

```
(temp + species)^2
```

Our focus, when seeing this, is that there are two predictors and the model should contain their main effects and the two-way interactions. However, this formula also implies that, since `species` is a factor, it should also create indicator variable columns for this predictor (see Chapter 8) and multiply those columns by the `temp` column to create the interactions. This transformation represents our second bullet point on encoding; the formula also defines how each column is encoded and can create additional columns that are not in the original data.

 This is an important point that will come up multiple times in this text, especially when we discuss more complex feature engineering in Chapter 8 and beyond. The formula in R has some limitations, and our approaches to overcoming them contend with all three aspects.

Why Tidiness Is Important for Modeling

One of the strengths of R is that it encourages developers to create a user interface that fits their needs. As an example, here are three common methods for creating a scatterplot of two numeric variables in a data frame called `plot_data`:

```
plot(plot_data$x, plot_data$y)

library(lattice)
xyplot(y ~ x, data = plot_data)

library(ggplot2)
ggplot(plot_data, aes(x = x, y = y)) + geom_point()
```

In these three cases, separate groups of developers devised three distinct interfaces for the same task. Each has advantages and disadvantages.

In comparison, the *Python Developer's Guide* espouses the notion that, when approaching a problem, "There should be one—and preferably only one—obvious way to do it."

R is quite different from Python in this respect. An advantage of R's diversity of interfaces is that it can evolve over time and fit different needs for different users.

Unfortunately, some of the syntactical diversity is due to a focus on the needs of the person *developing* the code instead of the needs of the person *using* the code. Inconsistencies among packages can be a stumbling block for R users.

Suppose your modeling project has an outcome with two classes. There are a variety of statistical and machine learning models you could choose from. In order to produce a class probability estimate for each sample, it is common for a model function to have a corresponding `predict()` method. However, there is significant heterogeneity in the argument values used by those methods to make class probability predictions; this heterogeneity can be difficult for even experienced users to navigate. A sampling of these argument values for different models is shown in Table 3-1.

Table 3-1. Heterogeneous argument names for different modeling functions.

Function	Package	Code
`lda()`	MASS	`predict(object)`
`glm()`	stats	`predict(object, type = "response")`
`gbm()`	gbm	`predict(object, type = "response", n.trees)`
`mda()`	mda	`predict(object, type = "posterior")`
`rpart()`	rpart	`predict(object, type = "prob")`
various	RWeka	`predict(object, type = "probability")`
`logitboost()`	LogitBoost	`predict(object, type = "raw", nIter)`
`pamr.train()`	pamr	`pamr.predict(object, type = "posterior")`

Note that the last example has a custom function to make predictions instead of using the more common `predict()` interface (the generic `predict()` method). This lack of consistency is a barrier to day-to-day usage of R for modeling.

As another example of unpredictability, the R language has conventions for missing data that are handled inconsistently. The general rule is that missing data propagate more missing data; the average of a set of values with a missing data point is itself missing and so on. When models make predictions, the vast majority require all of the predictors to have complete values. There are several options baked into R at this point with the generic function `na.action()`. This sets the policy for how a function should behave if there are missing values. The two most common policies are `na.fail()` and `na.omit()`. The former produces an error if missing data are present while the latter removes the missing data prior to calculations by case-wise deletion. From our previous example:

```
# Add a missing value to the prediction set
new_values$temp[1] <- NA

# The predict method for `lm` defaults to `na.pass`:
predict(main_effect_fit, new_values)
#>    1     2     3     4     5     6
#>   NA 50.43 54.04 57.64 61.24 64.84

# Alternatively
```

```
predict(main_effect_fit, new_values, na.action = na.fail)
#> Error in na.fail.default(structure(list(temp = c(NA, 16L, 17L, 18L, 19L, ...

predict(main_effect_fit, new_values, na.action = na.omit)
#>     2     3     4     5     6
#> 50.43 54.04 57.64 61.24 64.84
```

From a user's point of view, na.omit() can be problematic. In our example, new_values has six rows but only five would be returned with na.omit(). To adjust for this, the user would have to determine which row had the missing value and interleave a missing value in the appropriate place if the predictions were merged into new_values.[2] While it is rare that a prediction function uses na.omit() as its missing data policy, this does occur. Users who have determined this as the cause of an error in their code find it quite memorable.

To resolve the usage issues described here, the tidymodels packages have a set of design goals. Most of the tidymodels design goals fall under the existing rubric of "Design for Humans" from the tidyverse (Wickham et al. 2019), but with specific applications for modeling code. There are a few additional tidymodels design goals that complement those of the tidyverse; for example:

- R has excellent capabilities for object-oriented programming, and we use this in lieu of creating new function names (such as a hypothetical new predict _samples() function).

- *Sensible defaults* are very important. Also, functions should have no default for arguments when it is more appropriate to force the user to make a choice (e.g., the file name argument for read_csv()).

- Similarly, argument values whose default can be derived from the data should be. For example, for glm() the family argument could check the type of data in the outcome and, if no family was given, a default could be determined internally.

- Functions should take the *data structures that users have* as opposed to the data structure that developers want. For example, a model function's only interface should not be constrained to matrices. Frequently, users will have nonnumeric predictors such as factors.

Many of these ideas are described in the tidymodels guidelines for model implementation (*https://oreil.ly/qdshy*). In subsequent chapters, we will illustrate examples of existing issues, along with their solutions.

---

2 A base R policy called na.exclude() does exactly this.

A few existing R packages provide a unified interface to harmonize these heterogeneous modeling APIs, such as caret and mlr. The tidymodels framework is similar to these in adopting a unification of the function interface, as well as enforcing consistency in the function names and return values. It is different in its opinionated design goals and modeling implementation, as discussed in detail throughout this book.

The `broom::tidy()` function, which we use throughout this book, is another tool for standardizing the structure of R objects. It can return many types of R objects in a more usable format. For example, suppose that predictors are being screened based on their correlation to the outcome column. Using `purrr::map()`, the results from `cor.test()` can be returned in a list for each predictor:

```
corr_res <- map(mtcars %>% select(-mpg), cor.test, y = mtcars$mpg)

# The first of ten results in the vector:
corr_res[[1]]
#>
#>   Pearson's product-moment correlation
#>
#> data:  .x[[i]] and mtcars$mpg
#> t = -8.9, df = 30, p-value = 6e-10
#> alternative hypothesis: true correlation is not equal to 0
#> 95 percent confidence interval:
#>   -0.9258 -0.7163
#> sample estimates:
#>      cor
#> -0.8522
```

If we want to use these results in a plot, the standard format of hypothesis test results is not very useful. The `tidy()` method can return this as a tibble with standardized names:

```
library(broom)

tidy(corr_res[[1]])
#> # A tibble: 1 × 8
#>   estimate statistic  p.value parameter conf.low conf.high method     alternative
#>      <dbl>     <dbl>    <dbl>     <int>    <dbl>     <dbl> <chr>      <chr>
#> 1   -0.852     -8.92 6.11e-10        30   -0.926    -0.716 Pearson's pr… two.sided
```

These results can be "stacked" and added to a `ggplot()`, as shown in Figure 3-3:

```
corr_res %>%
  # Convert each to a tidy format; `map_dfr()` stacks the data frames
  map_dfr(tidy, .id = "predictor") %>%
  ggplot(aes(x = fct_reorder(predictor, estimate))) +
  geom_point(aes(y = estimate)) +
  geom_errorbar(aes(ymin = conf.low, ymax = conf.high), width = .1) +
  labs(x = NULL, y = "Correlation with mpg")
```

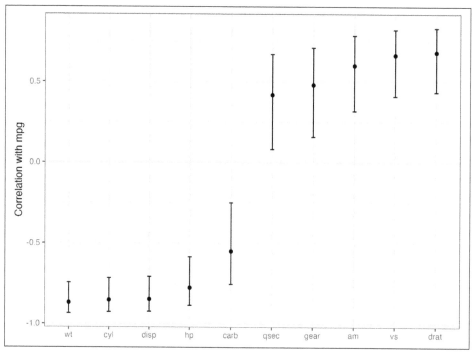

Figure 3-3. Correlations (and 95% confidence intervals) between predictors and the outcome in the mtcars *data set.*

Creating such a plot is possible using core R language functions, but automatically reformatting the results makes for more concise code with less potential for errors.

Combining Base R Models and the Tidyverse

R modeling functions from the core language or other R packages can be used in conjunction with the tidyverse, especially with the dplyr, purrr, and tidyr packages. For example, if we wanted to fit separate models for each cricket species, we could first break out the cricket data by this column using dplyr::group_nest():

```
split_by_species <-
  crickets %>%
  group_nest(species)
split_by_species
#> # A tibble: 2 × 2
#>   species                        data
#>   <fct>              <list<tibble[,2]>>
#> 1 O. exclamationis            [14 × 2]
#> 2 O. niveus                   [17 × 2]
```

The `data` column contains the `rate` and `temp` columns from `crickets` in a *list column*. From this, the `purrr::map()` function can create individual models for each species:

```
model_by_species <-
  split_by_species %>%
  mutate(model = map(data, ~ lm(rate ~ temp, data = .x)))
model_by_species
#> # A tibble: 2 × 3
#>   species                          data model
#>   <fct>              <list<tibble[,2]>> <list>
#> 1 O. exclamationis            [14 × 2] <lm>
#> 2 O. niveus                   [17 × 2] <lm>
```

To collect the coefficients for each of these models, use `broom::tidy()` to convert them to a consistent data frame format so that they can be unnested:

```
model_by_species %>%
  mutate(coef = map(model, tidy)) %>%
  select(species, coef) %>%
  unnest(cols = c(coef))
#> # A tibble: 4 × 6
#>   species          term        estimate std.error statistic  p.value
#>   <fct>            <chr>          <dbl>     <dbl>     <dbl>    <dbl>
#> 1 O. exclamationis (Intercept)   -11.0      4.77     -2.32 3.90e- 2
#> 2 O. exclamationis temp           3.75     0.184     20.4  1.10e-10
#> 3 O. niveus        (Intercept)   -15.4      2.35     -6.56 9.07e- 6
#> 4 O. niveus        temp           3.52     0.105     33.6  1.57e-15
```

 List columns can be very powerful in modeling projects. List columns provide containers for any type of R objects, from a fitted model itself to the important data frame structure.

The tidymodels Metapackage

The tidyverse (Chapter 2) is designed as a set of modular R packages, each with a fairly narrow scope. The tidymodels framework follows a similar design. For example, the rsample package focuses on data splitting and resampling. Although resampling methods are critical to other activities of modeling (e.g., measuring performance), they reside in a single package, and performance metrics are contained in a different, separate package, named yardstick. There are many benefits to adopting this philosophy of modular packages, from less bloated model deployment to smoother package maintenance.

The downside to this philosophy is that there are a lot of packages in the tidymodels framework. To compensate for this, the tidymodels *package* (which you can think of as a metapackage like the tidyverse package) loads a core set of tidymodels and tidyverse packages. Loading the package shows which packages are attached:

```
library(tidymodels)
#> — Attaching packages ───────────────────────────── tidymodels 0.2.0 —
#> ✓ broom       0.8.0      ✓ recipes     0.2.0
#> ✓ dials       0.1.1      ✓ rsample     0.1.1
#> ✓ dplyr       1.0.8      ✓ tibble      3.1.6
#> ✓ ggplot2     3.3.5      ✓ tidyr       1.2.0
#> ✓ infer       1.0.0      ✓ tune        0.2.0
#> ✓ modeldata   0.1.1      ✓ workflows   0.2.6
#> ✓ parsnip     0.2.1      ✓ workflowsets 0.2.1
#> ✓ purrr       0.3.4      ✓ yardstick   0.0.9
#> — Conflicts ──────────────────────────── tidymodels_conflicts() —
#> ✗ purrr::discard() masks scales::discard()
#> ✗ dplyr::filter()  masks stats::filter()
#> ✗ dplyr::lag()     masks stats::lag()
#> ✗ recipes::step()  masks stats::step()
#> • Learn how to get started at https://www.tidymodels.org/start
```

If you have used the tidyverse, you'll notice some familiar names as a few tidyverse packages, such as dplyr and ggplot2, are loaded together with the tidymodels packages. We've already said that the tidymodels framework applies tidyverse principles to modeling, but the tidymodels framework also literally builds on some of the most fundamental tidyverse packages such as these.

Loading the metapackage also shows if there are function naming conflicts with previously loaded packages. As an example of a naming conflict, before loading tidymodels, invoking the `filter()` function will execute the function in the stats package. After loading tidymodels, it will execute the dplyr function of the same name.

There are a few ways to handle naming conflicts. The function can be called with its namespace (e.g., `stats::filter()`). This is not a bad practice, but it does make the code less readable.

Another option is to use the conflicted package. We can set a rule that remains in effect until the end of the R session to ensure that one specific function will always run if no namespace is given in the code. As an example, if we prefer the dplyr version of the previous function:

```
library(conflicted)
conflict_prefer("filter", winner = "dplyr")
```

For convenience, tidymodels contains a function that captures most of the common naming conflicts that we might encounter:

```
tidymodels_prefer(quiet = FALSE)
#> [conflicted] Will prefer dplyr::filter over any other package
#> [conflicted] Will prefer dplyr::select over any other package
#> [conflicted] Will prefer dplyr::slice over any other package
#> [conflicted] Will prefer dplyr::rename over any other package
#> [conflicted] Will prefer dials::neighbors over any other package
#> [conflicted] Will prefer parsnip::fit over any other package
#> [conflicted] Will prefer parsnip::bart over any other package
#> [conflicted] Will prefer parsnip::pls over any other package
#> [conflicted] Will prefer purrr::map over any other package
```

```
#> [conflicted] Will prefer recipes::step over any other package
#> [conflicted] Will prefer themis::step_downsample over any other package
#> [conflicted] Will prefer themis::step_upsample over any other package
#> [conflicted] Will prefer tune::tune over any other package
#> [conflicted] Will prefer yardstick::precision over any other package
#> [conflicted] Will prefer yardstick::recall over any other package
#> [conflicted] Will prefer yardstick::spec over any other package
#> — Conflicts ————————————————————————— tidymodels_prefer() —
```

 Be aware that using this function opts you into using `conflicted::conflict_prefer()` for all namespace conflicts, making every conflict an error and forcing you to choose which function to use. The function `tidymodels::tidymodels_prefer()` handles the most common conflicts from tidymodels functions, but you will need to handle other conflicts in your R session yourself.

Chapter Summary

This chapter reviewed core R language conventions for creating and using models that are an important foundation for the rest of this book. The formula operator is an expressive and important aspect of fitting models in R and often serves multiple purposes in non-tidymodels functions. Traditional R approaches to modeling have some limitations, especially when it comes to fluently handling and visualizing model output. The tidymodels metapackage applies tidyverse design philosophy to modeling packages.

PART II
Modeling Basics

## CHAPTER 4

The Ames Housing Data

In this chapter, we'll introduce the Ames housing data set (De Cock 2011), which we will use in modeling examples throughout this book. Exploratory data analysis, like what we walk through in this chapter, is an important first step in building a reliable model. The data set contains information on 2,930 properties in Ames, Iowa, including columns related to:

- House characteristics (bedrooms, garage, fireplace, pool, porch, etc.)
- Location (neighborhood)
- Lot information (zoning, shape, size, etc.)
- Ratings of condition and quality
- Sale price

The raw housing data are provided in De Cock (2011), but in our analyses in this book, we use a transformed version available in the modeldata package. This version has several changes and improvements to the data (*https://oreil.ly/OSIQ0*). For example, the longitude and latitude values have been determined for each property. Also, some columns were modified to be more analysis ready. For example:

- In the raw data, if a house did not have a particular feature, it was implicitly encoded as missing. For example, 2,732 properties did not have an alleyway. Instead of leaving these as missing, they were relabeled in the transformed version to indicate that no alley was available.
- The categorical predictors were converted to R's factor data type. While both the tidyverse and base R have moved away from importing data as factors by default, this data type is a better approach for storing qualitative data for modeling than simple strings.

- We removed a set of quality descriptors for each house since they are more like outcomes than predictors.

To load the data:

```
library(modeldata) # This is also loaded by the tidymodels package
data(ames)

# or, in one line:
data(ames, package = "modeldata")

dim(ames)
#> [1] 2930   74
```

Figure 4-1 shows the locations of the properties in Ames. The locations will be revisited in the next section.

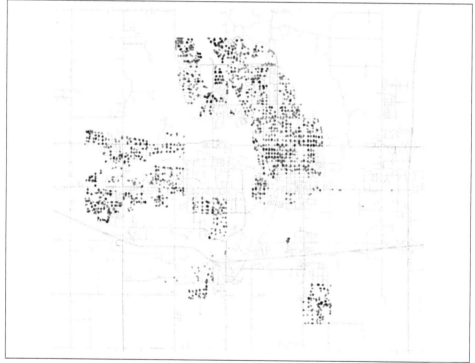

Figure 4-1. Property locations in Ames, Iowa.

The void of data points in the center of Ames corresponds to Iowa State University.

Our modeling goal is to predict the sale price of a house based on other information we have, such as its characteristics and location.

Exploring Features of Homes in Ames

Let's start our exploratory data analysis by focusing on the outcome we want to predict: the last sale price of the house (in USD). We can create a histogram to see the distribution of sale prices in Figure 4-2:

```
library(tidymodels)
tidymodels_prefer()

ggplot(ames, aes(x = Sale_Price)) +
  geom_histogram(bins = 50, col= "white")
```

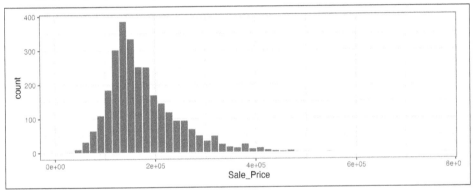

Figure 4-2. Sale prices of houses in Ames, Iowa.

This plot shows us that the data are right-skewed; there are more inexpensive houses than expensive ones. The median sale price was \$160,000, and the most expensive house was \$755,000. When modeling this outcome, a strong argument can be made that the price should be log-transformed. The advantages of this type of transformation are that no houses would be predicted with negative sale prices and that errors in predicting expensive houses will not have an undue influence on the model. Also, from a statistical perspective, a logarithmic transform may also stabilize the variance in a way that makes inference more legitimate. We now can use similar steps to visualize the transformed data, shown in Figure 4-3:

```
ggplot(ames, aes(x = Sale_Price)) +
  geom_histogram(bins = 50, col= "white") +
  scale_x_log10()
```

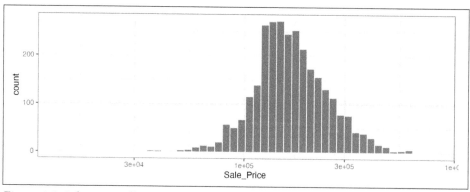

Figure 4-3. Sale prices of houses in Ames, Iowa after a log (base 10) transformation.

While not perfect, this will likely result in better models than using the untransformed data, for the reasons just outlined.

> The disadvantages of transforming the outcome mostly relate to interpretation of model results.

The units of the model coefficients might be more difficult to interpret, as will measures of performance. For example, the *root mean squared error* (RMSE) is a common performance metric used in regression models. It uses the difference between the observed and predicted values in its calculations. If the sale price is on the log scale, these differences (i.e., the residuals) are also on the log scale. It can be difficult to understand the quality of a model whose RMSE is 0.15 on such a log scale.

Despite these drawbacks, the models used in this book use the log transformation for this outcome. *From this point on*, the outcome column is prelogged in the `ames` data frame:

```
ames <- ames %>% mutate(Sale_Price = log10(Sale_Price))
```

Another important aspect of these data for our modeling are their geographic locations. This spatial information is contained in the data in two ways: a qualitative `Neighborhood` label as well as quantitative longitude and latitude data. To visualize the spatial information, Figure 4-4 duplicates the data from Figure 4-1 with convex hulls around the data from each neighborhood.

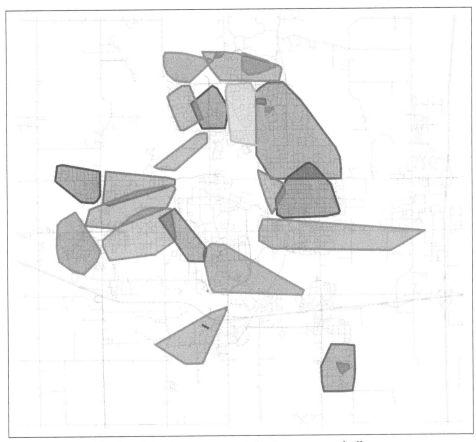

Figure 4-4. Neighborhoods in Ames represented using a convex hull.

We can see a few noticeable patterns. First, there is a void of data points in the center of Ames. This corresponds to the campus of Iowa State University where there are no residential houses. Second, while there are a number of adjacent neighborhoods, others are geographically isolated. For example, as Figure 4-5 shows, Timberland is located apart from almost all other neighborhoods.

Figure 4-6 visualizes how the Meadow Village neighborhood in southwest Ames is like an island of properties inside the sea of properties that make up the Mitchell neighborhood.

Figure 4-5. Locations of homes in Timberland.

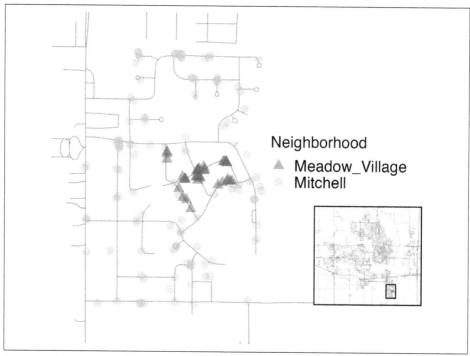

Figure 4-6. Locations of homes in Meadow Village and Mitchell.

A detailed inspection of the map also shows that the neighborhood labels are not completely reliable. For example, Figure 4-7 shows some properties labeled as being in Northridge that are surrounded by homes in the adjacent Somerset neighborhood.

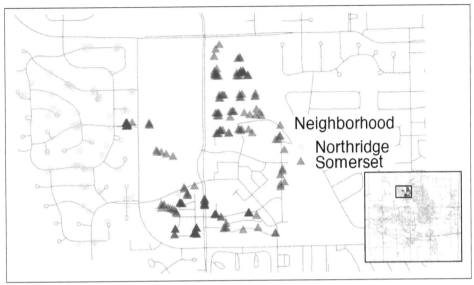

Figure 4-7. Locations of homes in Somerset and Northridge.

Also, there are 10 isolated homes labeled as being in Crawford that, as you can see in Figure 4-8, are not close to the majority of the other homes in that neighborhood.

Figure 4-8. Locations of homes in Crawford.

Also notable is the "Iowa Department of Transportation (DOT) and Rail Road" neighborhood adjacent to the main road on the east side of Ames, shown in Figure 4-9. There are several clusters of homes within this neighborhood as well as some longitudinal outliers; the two homes farthest east are isolated from the other locations.

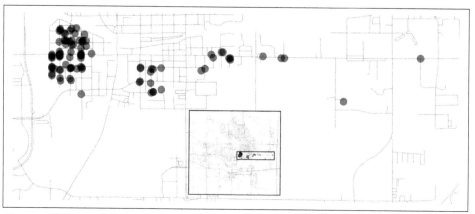

Figure 4-9. Homes labeled as Iowa Department of Transportation (DOT) and Rail Road.

As described in Chapter 1, it is critical to conduct exploratory data analysis prior to beginning any modeling. These housing data have characteristics that present interesting challenges about how the data should be processed and modeled. We describe many of these in later chapters, such as Chapter 17. Some basic questions that could be examined during this exploratory stage include:

- Is there anything odd or noticeable about the distributions of the individual predictors? Is there much skewness or any pathological distributions?
- Are there high correlations between predictors? For example, there are multiple predictors related to house size. Are some redundant?
- Are there associations between predictors and the outcomes?

Many of these questions will be revisited as these data are used in upcoming examples.

Chapter Summary

This chapter introduced the Ames housing data set and investigated some of its characteristics. This data set will be used in later chapters to demonstrate tidymodels syntax. Exploratory data analysis like this is an essential component of any modeling project; EDA uncovers information that contributes to better modeling practice.

The important code for preparing the Ames data set that we will carry forward into subsequent chapters is:

```
library(tidymodels)
data(ames)
ames <- ames %>% mutate(Sale_Price = log10(Sale_Price))
```

## CHAPTER 5

Spending Our Data

There are several steps to creating a useful model, including parameter estimation, model selection and tuning, and performance assessment. At the start of a new project, there is usually an initial finite pool of data available for all these tasks, which we can think of as an available data budget. How should the data be applied to different steps or tasks? The idea of *data spending* is an important first consideration when modeling, especially as it relates to empirical validation.

 When data are reused for multiple tasks, instead of being carefully "spent" from the finite data budget, certain risks increase, such as the risk of accentuating bias or compounding effects from methodological errors.

When there are copious amounts of data available, a smart strategy is to allocate specific subsets of data for different tasks, as opposed to allocating the largest possible amount (or even all) to the model parameter estimation only. For example, one possible strategy (when both data and predictors are abundant) is to spend a specific subset of data to determine which predictors are informative, before considering parameter estimation at all. If the initial pool of data available is not huge, there will be some overlap in how and when our data is spent or allocated, and a solid methodology for data spending is important.

This chapter demonstrates the basics of *splitting* (i.e., creating a data budget) for our initial pool of samples for different purposes.

Common Methods for Splitting Data

The primary approach for empirical model validation is to split the existing pool of data into two distinct sets, the training set and the test set. One portion of the data is used to develop and optimize the model. This *training set* is usually the majority of the data. These data are a sandbox for model building where different models can be fit, feature engineering strategies are investigated, and so on. As modeling practitioners we spend the vast majority of the modeling process using the training set as the substrate to develop the model.

The other portion of the data is placed into the *test set*. This is held in reserve until one or two models are chosen as the methods most likely to succeed. The test set is then used as the final arbiter to determine the efficacy of the model. It is critical to look at the test set only once; otherwise, it becomes part of the modeling process.

How should we conduct this split of the data? The answer depends on the context.

Suppose we allocate 80% of the data to the training set and the remaining 20% for testing. The most common method is to use simple random sampling. The rsample package (*https://oreil.ly/a9sWI*) has tools for making data splits such as this; the function `initial_split()` was created for this purpose. It takes the data frame as an argument as well as the proportion to be placed into training. We can set the split using the data frame produced by the code snippet from the summary at the end of Chapter 4:

```
library(tidymodels)
tidymodels_prefer()

# Set the random number stream using `set.seed()` so that the results can be
# reproduced later.
set.seed(501)

# Save the split information for an 80/20 split of the data
ames_split <- initial_split(ames, prop = 0.80)
ames_split
#> <Analysis/Assess/Total>
#> <2344/586/2930>
```

The printed information denotes the amount of data in the training set ($n = 2,344$), the amount in the test set ($n = 586$), and the size of the original pool of samples ($n = 2,930$).

The object `ames_split` is an `rsplit` object and contains only the partitioning information; to get the resulting data sets, we apply two more functions:

```
ames_train <- training(ames_split)
ames_test  <- testing(ames_split)

dim(ames_train)
#> [1] 2344    74
```

These objects are data frames with the same columns as the original data but only the appropriate rows for each set.

Simple random sampling is appropriate in many cases, but there are exceptions. When there is a dramatic *class imbalance* in classification problems, one class occurs much less frequently than another. Using a simple random sample may haphazardly allocate these infrequent samples disproportionately into the training or test set. To avoid this, *stratified sampling* can be used. The training/test split is conducted separately within each class, and then these subsamples are combined into the overall training and test set. For regression problems, the outcome data can be artificially binned into quartiles, and then stratified sampling can be conducted four separate times. This is an effective method for keeping the distributions of the outcome similar between the training and test sets. The distribution of the sale price outcome for the Ames housing data is shown in Figure 5-1.

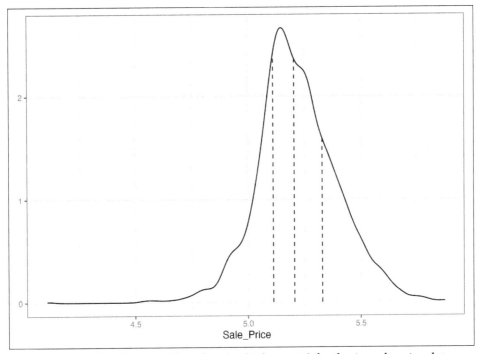

Figure 5-1. The distribution of the sale price (in log units) for the Ames housing data. The vertical lines indicate the quartiles of the data.

As discussed in Chapter 4, the sale price distribution is right-skewed, with proportionally more inexpensive houses than expensive houses on either side of the center of the distribution. The worry here with simple splitting is that the more expensive houses would not be well represented in the training set; this would increase the risk that our model would be ineffective at predicting the price for such properties. The dotted vertical lines in Figure 5-1 indicate the four quartiles for these data. A stratified random sample would conduct the 80/20 split within each of these data subsets and then pool the results. In rsample, this is achieved using the strata argument:

```
set.seed(502)
ames_split <- initial_split(ames, prop = 0.80, strata = Sale_Price)
ames_train <- training(ames_split)
ames_test  <- testing(ames_split)

dim(ames_train)
#> [1] 2342    74
```

Only a single column can be used for stratification.

There is very little downside to using stratified sampling.

Are there situations when random sampling is not the best choice? One case is when the data have a significant time component, such as time series data. Here, it is more common to use the most recent data as the test set. The rsample package contains a function called initial_time_split() that is very similar to initial_split(). Instead of using random sampling, the prop argument denotes what proportion of the first part of the data should be used as the training set; the function assumes that the data have been presorted in an appropriate order.

The proportion of data that should be allocated for splitting is highly dependent on the context of the problem at hand. Too little data in the training set hampers the model's ability to find appropriate parameter estimates. Conversely, too little data in the test set lowers the quality of the performance estimates. Parts of the statistics community eschew test sets in general because they believe all of the data should be used for parameter estimation. While there is merit to this argument, it is good modeling practice to have an unbiased set of observations as the final arbiter of model quality. A test set should be avoided only when the data set is pathologically small.

What About a Validation Set?

When describing the goals of data splitting, we singled out the test set as the data that should be used to properly evaluate model performance on the final model(s). This begs the question: "How can we tell what is best if we don't measure performance until the test set?"

It is common to hear about *validation sets* as an answer to this question, especially in the neural network and deep learning literature. During the early days of neural networks, researchers realized that measuring performance by repredicting the training set samples led to results that were overly optimistic (significantly, unrealistically so). This led to models that overfit, meaning that they performed very well on the training set but poorly on the test set.[1] To combat this issue, a small validation data set was held back and used to measure performance as the network was trained. Once the validation set error rate began to rise, the training would be halted. In other words, the validation set was a means to get a rough sense of how well the model performed prior to the test set.

 Whether validation sets are a subset of the training set or a third allocation in the initial split of the data largely comes down to semantics.

Validation sets are discussed more in Chapter 10 as a special case of *resampling* methods that are used on the training set.

Multilevel Data

With the Ames housing data, a property is considered to be the *independent experimental unit*. It is safe to assume that, statistically, the data from a property are independent of other properties. For other applications, that is not always the case:

- For longitudinal data, for example, the same independent experimental unit can be measured over multiple time points. An example would be a human subject in a medical trial.
- A batch of manufactured product might also be considered the independent experimental unit. In repeated measures designs, replicate data points from a batch are collected at multiple times.

---

1 This is discussed in much greater detail in Chapter 12.

- Johnson et al. (2018) report an experiment in which different trees were sampled across the top and bottom portions of a stem. Here, the tree is the experimental unit and the data hierarchy is sample within stem position within tree.

Chapter 9 of Kuhn and Johnson (2020) (*https://oreil.ly/bbEW1*) contains other examples.

In these situations, the data set will have multiple rows per experimental unit. Simple resampling across rows would lead to some data within an experimental unit being in the training set and others in the test set. Data splitting should occur at the independent experimental unit level of the data. For example, to produce an 80/20 split of the Ames housing data set, 80% of the properties should be allocated for the training set.

Other Considerations for a Data Budget

When deciding how to spend the data available to you, keep a few more things in mind. First, it is critical to quarantine the test set from any model building activities. As you read this book, notice which data are exposed to the model at any given time.

 The problem of *information leakage* occurs when data outside of the training set are used in the modeling process.

For example, in a machine learning competition, the test set data might be provided without the true outcome values so that the model can be scored and ranked. One potential method for improving the score might be to fit the model using the training set points that are most similar to the test set values. While the test set isn't directly used to fit the model, it still has a heavy influence. In general, this technique is highly problematic because it reduces the *generalization error* of the model to optimize performance on a specific data set. There are more subtle ways that the test set data can be used during training. Keeping the training data in a separate data frame from the test set is one small check to make sure that information leakage does not occur by accident.

Second, techniques to subsample the training set can mitigate specific issues (e.g., class imbalances). This is a valid and common technique that deliberately results in the training set data diverging from the population from which the data were drawn. It is critical that the test set continues to mirror what the model would encounter in the wild. In other words, the test set should always resemble new data that will be given to the model.

Next, at the beginning of this chapter, we warned about using the same data for different tasks. Chapter 10 will discuss solid, data-driven methodologies for data usage that will reduce the risks related to bias, overfitting, and other issues. Many of these methods apply the data-splitting tools introduced in this chapter.

Finally, the considerations in this chapter apply to developing and choosing a reliable model, the main topic of this book. When training a final chosen model for production, after ascertaining the expected performance on new data, practitioners often use all available data for better parameter estimation.

Chapter Summary

Data splitting is the fundamental strategy for empirical validation of models. Even in the era of unrestrained data collection, a typical modeling project has a limited amount of appropriate data, and wise spending of a project's data is necessary. In this chapter, we discussed several strategies for partitioning the data into distinct groups for modeling and evaluation.

At this checkpoint, the important code snippets for preparing and splitting are:

```
library(tidymodels)
data(ames)
ames <- ames %>% mutate(Sale_Price = log10(Sale_Price))

set.seed(502)
ames_split <- initial_split(ames, prop = 0.80, strata = Sale_Price)
ames_train <- training(ames_split)
ames_test  <- testing(ames_split)
```

## CHAPTER 6

Fitting Models with parsnip

The parsnip package is one of the R packages that is part of the tidymodels metapackage. It provides a fluent and standardized interface for a variety of different models. In this chapter, we give some motivation for why a common interface is beneficial for understanding and building models in practice and show how to use the parsnip package.

Specifically, we will focus on how to `fit()` and `predict()` directly with a parsnip object, which may be a good fit for some straightforward modeling problems. Chapter 7 illustrates a better approach for many modeling tasks by combining models and preprocessors together into something called a `workflow` object.

Create a Model

Once the data have been encoded in a format ready for a modeling algorithm, such as a numeric matrix, they can be used in the model building process.

Suppose that a linear regression model was our initial choice. This is equivalent to specifying that the outcome data are numeric and that the predictors are related to the outcome in terms of simple slopes and intercepts:

$$y_i = \beta_0 + \beta_1 x_{1i} + \dots + \beta_p x_{pi}$$

A variety of methods can be used to estimate the model parameters:

- *Ordinary linear regression* uses the traditional method of least squares to solve for the model parameters.

- *Regularized linear regression* adds a penalty to the least squares method to encourage simplicity by removing predictors and/or shrinking their coefficients toward zero. This can be executed using Bayesian or non-Bayesian techniques.

In R, the stats package can be used for the first case. The syntax for linear regression using the function lm( ) is:

```
model <- lm(formula, data, ...)
```

where ... symbolizes other options to pass to lm( ). The function does *not* have an x/y interface, where we might pass in our outcome as y and our predictors as x.

To estimate with regularization, the second case, a Bayesian model can be fit using the rstanarm package:

```
model <- stan_glm(formula, data, family = "gaussian", ...)
```

In this case, the other options passed via ... would include arguments for the prior distributions of the parameters as well as specifics about the numerical aspects of the model. As with lm( ), only the formula interface is available.

A popular non-Bayesian approach to regularized regression is the glmnet model (Friedman, Hastie, and Tibshirani 2010). Its syntax is:

```
model <- glmnet(x = matrix, y = vector, family = "gaussian", ...)
```

In this case, the predictor data must already be formatted into a numeric matrix; there is only an x/y method and no formula method.

Note that these interfaces are heterogeneous in either how the data are passed to the model function or in terms of their arguments. The first issue is that, to fit models across different packages, the data must be formatted in different ways. lm( ) and stan_glm() only have formula interfaces whereas glmnet() does not. For other types of models, the interfaces may be even more disparate. For a person trying to do data analysis, these differences require the memorization of each package's syntax and can be very frustrating.

For tidymodels, the approach to specifying a model is intended to be more unified:

Specify the type of model based on its mathematical structure
   Such as linear regression, random forest, KNN, etc.Most often this reflects the software package that should be used, like Stan or glmnet. These are models in their own right, and parsnip provides consistent interfaces by using these as engines for modeling.

When required, declare the mode of the model

The mode reflects the type of prediction outcome. For numeric outcomes, the mode is regression; for qualitative outcomes, it is classification.[1] If a model algorithm can only address one type of prediction outcome, such as linear regression, the mode is already set.

These specifications are built without referencing the data. For example, for the three cases we outlined:

```
library(tidymodels)
tidymodels_prefer()

linear_reg() %>% set_engine("lm")
#> Linear Regression Model Specification (regression)
#>
#> Computational engine: lm

linear_reg() %>% set_engine("glmnet")
#> Linear Regression Model Specification (regression)
#>
#> Computational engine: glmnet

linear_reg() %>% set_engine("stan")
#> Linear Regression Model Specification (regression)
#>
#> Computational engine: stan
```

Once the details of the model have been specified, the model estimation can be done with either the `fit()` function (to use a formula) or the `fit_xy()` function (when your data are already preprocessed). The parsnip package allows the user to be indifferent to the interface of the underlying model; you can always use a formula even if the modeling package's function only has the x/y interface.

The `translate()` function can provide details on how parsnip converts the user's code to the package's syntax:

```
linear_reg() %>% set_engine("lm") %>% translate()
#> Linear Regression Model Specification (regression)
#>
#> Computational engine: lm
#>
#> Model fit template:
#> stats::lm(formula = missing_arg(), data = missing_arg(), weights = missing_arg())

linear_reg(penalty = 1) %>% set_engine("glmnet") %>% translate()
#> Linear Regression Model Specification (regression)
#>
#> Main Arguments:
#>   penalty = 1
#>
```

---

1 Note that parsnip constrains the outcome column of a classification model to be encoded as a *factor*; using binary numeric values will result in an error.

```
#> Computational engine: glmnet
#>
#> Model fit template:
#> glmnet::glmnet(x = missing_arg(), y = missing_arg(), weights = missing_arg(),
#>     family = "gaussian")

linear_reg() %>% set_engine("stan") %>% translate()
#> Linear Regression Model Specification (regression)
#>
#> Computational engine: stan
#>
#> Model fit template:
#> rstanarm::stan_glm(formula = missing_arg(), data = missing_arg(),
#>     weights = missing_arg(), family = stats::gaussian, refresh = 0)
```

Note that missing_arg() is just a placeholder for the data that has yet to be provided.

 We supplied a required penalty argument for the glmnet engine. Also, for the Stan and glmnet engines, the family argument was automatically added as a default. As will be shown later in this section, this option can be changed.

Let's walk through how to predict the sale price of houses in the Ames data as a function of only longitude and latitude:[2]

```
lm_model <-
  linear_reg() %>%
  set_engine("lm")

lm_form_fit <-
  lm_model %>%
  # Recall that Sale_Price has been pre-logged
  fit(Sale_Price ~ Longitude + Latitude, data = ames_train)

lm_xy_fit <-
  lm_model %>%
  fit_xy(
    x = ames_train %>% select(Longitude, Latitude),
    y = ames_train %>% pull(Sale_Price)
  )

lm_form_fit
#> parsnip model object
#>
```

---

2 What are the differences between fit() and fit_xy()? The fit_xy() function always passes the data as is to the underlying model function. It will not create dummy/indicator variables before doing so. When fit() is used with a model specification, this almost always means that dummy variables will be created from qualitative predictors. If the underlying function requires a matrix (like glmnet), it will make them. However, if the underlying function uses a formula, fit() just passes the formula to that function. We estimate that 99% of modeling functions using formulas make dummy variables. The other 1% include tree-based methods that do not require purely numeric predictors. See "How Does a workflow() Use the Formula?" on page 81 for more about using formulas in tidymodels.

```
#>
#> Call:
#> stats::lm(formula = Sale_Price ~ Longitude + Latitude, data = data)
#>
#> Coefficients:
#> (Intercept)     Longitude     Latitude
#>     -302.97        -2.07         2.71
lm_xy_fit
#> parsnip model object
#>
#>
#> Call:
#> stats::lm(formula = ..y ~ ., data = data)
#>
#> Coefficients:
#> (Intercept)     Longitude     Latitude
#>     -302.97        -2.07         2.71
```

Not only does parsnip enable a consistent model interface for different packages, it also provides consistency in the model arguments. It is common for different functions that fit the same model to have different argument names. Random forest model functions are a good example. Three commonly used arguments are the number of trees in the ensemble, the number of predictors to randomly sample with each split within a tree, and the number of data points required to make a split. For three different R packages implementing this algorithm, those arguments are shown in Table 6-1.

Table 6-1. Example argument names for different random forest functions

Argument type	ranger	randomForest	sparklyr
Number of sampled predictors	mtry	mtry	feature_subset_strategy
Number of trees	num.trees	ntree	num_trees
Number of data points to split	min.node.size	nodesize	min_instances_per_node

In an effort to make argument specification less painful, parsnip uses common argument names within and between packages. Table 6-2 shows, for random forests, what parsnip models use.

Table 6-2. Random forest argument names used by parsnip

Argument type	parsnip
Number of sampled predictors	mtry
Number of trees	trees
Number of data points to split	min_n

Admittedly, this is one more set of arguments to memorize. However, when other types of models have the same argument types, these names still apply. For example, boosted tree ensembles also create a large number of tree-based models, so trees is also used there, as is min_n, and so on.

Some of the original argument names can be fairly jargony. For example, to specify the amount of regularization to use in a glmnet model, the Greek letter lambda is used. While this mathematical notation is commonly used in the statistics literature, it is not obvious to many people what lambda represents (especially those who consume the model results). Since this is the penalty used in regularization, parsnip standardizes on the argument name penalty. Similarly, the number of neighbors in a KNN model is called neighbors instead of k. Our rule of thumb when standardizing argument names is:

> If a practitioner were to include these names in a plot or table, would the people viewing those results understand the name?

To understand how the parsnip argument names map to the original names, use the help file for the model (available via ?rand_forest) as well as the translate() function:

```
rand_forest(trees = 1000, min_n = 5) %>%
  set_engine("ranger") %>%
  set_mode("regression") %>%
  translate()
#> Random Forest Model Specification (regression)
#>
#> Main Arguments:
#>   trees = 1000
#>   min_n = 5
#>
#> Computational engine: ranger
#>
#> Model fit template:
#> ranger::ranger(x = missing_arg(), y = missing_arg(), case.weights = missing_arg(),
#>     num.trees = 1000, min.node.size = min_rows(~5, x), num.threads = 1,
#>     verbose = FALSE, seed = sample.int(10^5, 1))
```

Modeling functions in parsnip separate model arguments into two categories:

Main arguments
   More commonly used and tend to be available across engines

Engine arguments
   Either specific to a particular engine or used more rarely

For example, in the translation of the previous random forest code, the arguments num.threads, verbose, and seed were added by default. These arguments are specific to the ranger implementation of random forest models and wouldn't make sense as main arguments. Engine-specific arguments can be specified in set_engine(). For example, to have the ranger::ranger() function print out more information about the fit:

```
rand_forest(trees = 1000, min_n = 5) %>%
  set_engine("ranger", verbose = TRUE) %>%
  set_mode("regression")
#> Random Forest Model Specification (regression)
#>
#> Main Arguments:
#>   trees = 1000
#>   min_n = 5
#>
#> Engine-Specific Arguments:
#>   verbose = TRUE
#>
#> Computational engine: ranger
```

Use the Model Results

Once the model is created and fit, we can use the results in a variety of ways; we might want to plot, print, or otherwise examine the model output. Several quantities are stored in a parsnip model object, including the fitted model. This can be found in an element called fit, which can be returned using the extract_fit_engine() function:

```
lm_form_fit %>% extract_fit_engine()
#>
#> Call:
#> stats::lm(formula = Sale_Price ~ Longitude + Latitude, data = data)
#>
#> Coefficients:
#> (Intercept)    Longitude     Latitude
#>    -302.97        -2.07         2.71
```

Normal methods can be applied to this object, such as printing and plotting:

```
lm_form_fit %>% extract_fit_engine() %>% vcov()
#>             (Intercept) Longitude Latitude
#> (Intercept)     207.311   1.57466 -1.42397
#> Longitude         1.575   0.01655 -0.00060
#> Latitude         -1.424  -0.00060  0.03254
```

 Never pass the fit element of a parsnip model to a model prediction function, i.e., use predict(lm_form_fit) but *do not* use predict(lm_form_fit$fit). If the data were preprocessed in any way, incorrect predictions would be generated (sometimes, without errors). The underlying model's prediction function has no idea if any transformations have been made to the data prior to running the model. See "Make Predictions" on page 71 for more on making predictions.

One issue with some existing methods in base R is that the results are stored in a manner that may not be the most useful. For example, the summary() method for lm objects can be used to print the results of the model fit, including a table with parameter values, their uncertainty estimates, and p-values. These particular results can also be saved:

```
model_res <-
  lm_form_fit %>%
  extract_fit_engine() %>%
  summary()

# The model coefficient table is accessible via the `coef` method.
param_est <- coef(model_res)
class(param_est)
#> [1] "matrix" "array"
param_est
#>             Estimate Std. Error t value  Pr(>|t|)
#> (Intercept) -302.974    14.3983  -21.04 3.640e-90
#> Longitude     -2.075     0.1286  -16.13 1.395e-55
#> Latitude       2.710     0.1804   15.02 9.289e-49
```

There are a few things to notice about this result. First, the object is a numeric matrix. This data structure was mostly likely chosen since all of the calculated results are numeric and a matrix object is stored more efficiently than a data frame. This choice was probably made in the late 1970s when computational efficiency was extremely critical. Second, the nonnumeric data (the labels for the coefficients) are contained in the row names. Keeping the parameter labels as row names is very consistent with the conventions in the original S language.

A reasonable next step might be to create a visualization of the parameter values. To do this, it would be sensible to convert the parameter matrix to a data frame. We could add the row names as a column so that they can be used in a plot. However, notice that several of the existing matrix column names would not be valid R column names for ordinary data frames (e.g., `"Pr(>|t|)"`). Another complication is the consistency of the column names. For `lm` objects, the column for the test statistic is `"Pr(>|t|)"`, but for other models, a different test might be used and, as a result, the column name would be different (e.g., `"Pr(>|z|)"`) and the type of test would be encoded in the column name.

While these additional data formatting steps are not impossible to overcome, they are a hindrance, especially since they might be different for different types of models. The matrix is not a highly reusable data structure mostly because it constrains the data to be of a single type (e.g., numeric). Additionally, keeping some data in the dimension names is also problematic since those data must be extracted to be of general use.

As a solution, the broom package can convert many types of model objects to a tidy structure. For example, using the `tidy()` method on the linear model produces:

```
tidy(lm_form_fit)
#> # A tibble: 3 × 5
#>   term        estimate std.error statistic  p.value
#>   <chr>          <dbl>     <dbl>     <dbl>    <dbl>
#> 1 (Intercept)  -303.       14.4     -21.0 3.64e-90
```

```
#> 2 Longitude    -2.07    0.129   -16.1 1.40e-55
#> 3 Latitude      2.71    0.180    15.0 9.29e-49
```

The column names are standardized across models and do not contain any additional data (such as the type of statistical test). The data previously contained in the row names are now in a column called term. One important principle in the tidymodels ecosystem is that a function should return values that are *predictable, consistent,* and *unsurprising.*

Make Predictions

Another area where parsnip diverges from conventional R modeling functions is the format of values returned from predict(). For predictions, parsnip always conforms to the following rules:

1. The results are always a tibble.
2. The column names of the tibble are always predictable.
3. There are always as many rows in the tibble as there are in the input data set.

For example, when numeric data are predicted:

```
ames_test_small <- ames_test %>% slice(1:5)
predict(lm_form_fit, new_data = ames_test_small)
#> # A tibble: 5 × 1
#>    .pred
#>    <dbl>
#> 1  5.22
#> 2  5.21
#> 3  5.28
#> 4  5.27
#> 5  5.28
```

The row order of the predictions are always the same as the original data.

Why the leading dot in some of the column names? Some tidyverse and tidymodels arguments and return values contain periods. This is to protect against merging data with duplicate names. There are some data sets that contain predictors named pred!

These three rules make it easier to merge predictions with the original data:

```
ames_test_small %>%
  select(Sale_Price) %>%
  bind_cols(predict(lm_form_fit, ames_test_small)) %>%
  # Add 95% prediction intervals to the results:
  bind_cols(predict(lm_form_fit, ames_test_small, type = "pred_int"))
#> # A tibble: 5 × 4
#>    Sale_Price .pred .pred_lower .pred_upper
#>         <dbl> <dbl>       <dbl>       <dbl>
#> 1        5.02  5.22        4.91        5.54
```

```
#> 2     5.39  5.21      4.90       5.53
#> 3     5.28  5.28      4.97       5.60
#> 4     5.28  5.27      4.96       5.59
#> 5     5.28  5.28      4.97       5.60
```

The motivation for the first rule comes from some R packages producing dissimilar data types from prediction functions. For example, the ranger package is an excellent tool for computing random forest models. However, instead of returning a data frame or vector as output, it returns a specialized object that has multiple values embedded within it (including the predicted values). This is just one more step for the data analyst to work around in their scripts. As another example, the native glmnet model can return at least four different output types for predictions, depending on the model specifics and characteristics of the data. These are shown in Table 6-3.

Table 6-3. Different return values for glmnet prediction types

Type of prediction	Returns a:
Numeric	Numeric matrix
Class	character matrix
Probability (2 classes)	Numeric matrix (2nd level only)
Probability (3+ classes)	3D numeric array (all levels)

Additionally, the column names of the results contain coded values that map to a vector called lambda within the glmnet model object. This excellent statistical method can be discouraging to use in practice because of all of the special cases an analyst might encounter that require additional code to be useful.

For the second tidymodels prediction rule, the predictable column names for different types of predictions are shown in Table 6-4.

Table 6-4. The tidymodels mapping of prediction types and column names

Type value	Column name(s)
numeric	.pred
class	.pred_class
prob	.pred_{class levels}
conf_int	.pred_lower, .pred_upper
pred_int	.pred_lower, .pred_upper

The third rule regarding the number of rows in the output is critical. For example, if any rows of the new data contain missing values, the output will be padded with missing results for those rows. A main advantage of standardizing the model interface and prediction types in parsnip is that, when different models are used, the syntax is identical. Suppose that we used a decision tree to model the Ames data.

Outside of the model specification, there are no significant differences in the code pipeline:

```
tree_model <-
  decision_tree(min_n = 2) %>%
  set_engine("rpart") %>%
  set_mode("regression")

tree_fit <-
  tree_model %>%
  fit(Sale_Price ~ Longitude + Latitude, data = ames_train)

ames_test_small %>%
  select(Sale_Price) %>%
  bind_cols(predict(tree_fit, ames_test_small))
#> # A tibble: 5 × 2
#>   Sale_Price .pred
#>        <dbl> <dbl>
#> 1       5.02  5.15
#> 2       5.39  5.15
#> 3       5.28  5.32
#> 4       5.28  5.32
#> 5       5.28  5.32
```

This demonstrates the benefit of homogenizing the data analysis process and syntax across different models. It enables users to spend their time on the results and interpretation rather than having to focus on the syntactical differences between R packages.

parsnip-Extension Packages

The parsnip package itself contains interfaces to a number of models. However, for ease of package installation and maintenance, there are other tidymodels packages that have parsnip model definitions for other sets of models. The discrim package has model definitions for the set of classification techniques called discriminant analysis methods (such as linear or quadratic discriminant analysis). In this way, the package dependencies required for installing parsnip are reduced. A list of all of the models that can be used with parsnip (across different packages that are on CRAN) can be found on the tidymodels website (*https://oreil.ly/FB0BM*).

Creating Model Specifications

It may become tedious to write many model specifications, or to remember how to write the code to generate them. The parsnip package includes an RStudio addin (*https://oreil.ly/8qhDY*) that can help. Either choosing this addin from the *Addins* toolbar menu or running the code:

```
parsnip_addin()
```

will open a window in the Viewer panel of the RStudio IDE with a list of possible models for each model mode. These can be written to the source code panel.

The model list includes models from parsnip and parsnip-adjacent packages that are on CRAN.

Chapter Summary

This chapter introduced the parsnip package, which provides a common interface for models across R packages using a standard syntax. The interface and resulting objects have a predictable structure.

The code for modeling the Ames data that we will use moving forward is:

```
library(tidymodels)
data(ames)
ames <- mutate(ames, Sale_Price = log10(Sale_Price))

set.seed(502)
ames_split <- initial_split(ames, prop = 0.80, strata = Sale_Price)
ames_train <- training(ames_split)
ames_test  <- testing(ames_split)

lm_model <- linear_reg() %>% set_engine("lm")
```

## CHAPTER 7

A Model Workflow

In Chapter 6, we discussed the parsnip package, which can be used to define and fit the model. This chapter introduces a new concept called a *model workflow*. The purpose of this concept (and the corresponding tidymodels `workflow()` object) is to encapsulate the major pieces of the modeling process (discussed in Chapter 1). The workflow is important in two ways. First, using a workflow concept encourages good methodology since it is a single point of entry to the estimation components of a data analysis. Second, it enables the user to better organize projects. These two points are discussed in the following sections.

Where Does the Model Begin and End?

So far, when we have used the term "the model," we have meant a structural equation that relates some predictors to one or more outcomes. Let's consider again linear regression as an example. The outcome data are denoted as y_i, where there are $i = 1...n$ samples in the training set. Suppose that there are p predictors $x_{i1}, ..., x_{ip}$ that are used in the model. Linear regression produces the following model equation:

$$\widehat{y}_i = \widehat{\beta}_0 + \widehat{\beta}_1 x_{i1} + ... + \widehat{\beta}_p x_{ip}$$

While this is a linear model, it is linear only in the parameters. The predictors could be nonlinear terms (such as the log (x_i)).

The conventional way of thinking about the modeling process is that it only includes the model fit.

For some straightforward data sets, fitting the model itself may be the entire process. However, a variety of choices and additional steps often occur before the model is fit:

- While our example model has p predictors, it is common to start with more than p candidate predictors. Through exploratory data analysis or using domain knowledge, some of the predictors may be excluded from the analysis. In other cases, a feature selection algorithm may be used to make a data-driven choice for the minimum predictor set for the model.

- There are times when the value of an important predictor is missing. Rather than eliminating this sample from the data set, the missing value could be imputed using other values in the data. For example, if x_1 were missing but correlated with predictors x_2 and x_3, an imputation method could estimate the missing x_1 observation from the values of x_2 and x_3.

- It may be beneficial to transform the scale of a predictor. If there is not a priori information on what the new scale should be, we can estimate the proper scale using a statistical transformation technique, the existing data, and some optimization criterion. Other transformations, such as PCA, take groups of predictors and transform them into new features that are used as the predictors.

While these examples are related to steps that occur before the model fit, there may also be operations that occur after the model is created. When a classification model is created for which the outcome is binary (e.g., event and non-event), it is customary to use a 50% probability cutoff to create a discrete class prediction, also known as a hard prediction. For example, a classification model might estimate that the probability of an event was 62%. Using the typical default, the hard prediction would be event. However, the model may need to be more focused on reducing false positive results (i.e., where true nonevents are classified as events). One way to do this is to raise the cutoff from 50% to some greater value. This increases the level of evidence required to call a new sample an event. While this reduces the true positive rate (which is bad), it may have a more dramatic effect on reducing false positives. The choice of the cutoff value should be optimized using data. This is an example of a postprocessing step that has a significant effect on how well the model works, even though it is not contained in the model fitting step.

It is important to focus on the broader *modeling process*, instead of only fitting the specific model used to estimate parameters. This broader process includes any preprocessing steps, the model fit itself, as well as potential postprocessing activities. In this book, we will refer to this more comprehensive concept as the *model workflow* and highlight how to handle all its components to produce a final model equation.

In software such as Python or Spark, similar collections of steps are called *pipelines*. In tidymodels, the term *pipeline* already connotes a sequence of operations chained together with a pipe operator (such as %>% from magrittr or the newer native |>). Rather than using ambiguous terminology in this context, we call the sequence of computational operations related to modeling *workflows*.

Binding together the analytical components of data analysis is important for another reason. Future chapters show how to accurately measure performance, as well as how to optimize structural parameters (i.e., model tuning). To correctly quantify model performance on the training set, Chapter 10 advocates using resampling methods. To do this properly, no data-driven parts of the analysis should be excluded from validation. To this end, the workflow must include all significant estimation steps.

To illustrate, consider principal component analysis (PCA) signal extraction. We'll talk about this more in Chapter 8 as well as Chapter 16; PCA is a way to replace correlated predictors with new artificial features that are uncorrelated and capture most of the information in the original set. The new features could be used as the predictors, and least squares regression could be used to estimate the model parameters.

There are two ways of thinking about the model workflow. Figure 7-1 illustrates the *incorrect* method: to think of the PCA preprocessing step as *not being part of the modeling workflow*.

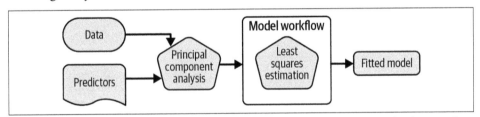

Figure 7-1. Incorrect mental model of where model estimation occurs in the data analysis process.

The fallacy here is that, although PCA does significant computations to produce the components, its operations are assumed to have no uncertainty associated with them. The PCA components are treated as *known*, and if they are not included in the model workflow, the effect of PCA could not be adequately measured.

Figure 7-2 shows an *appropriate* approach.

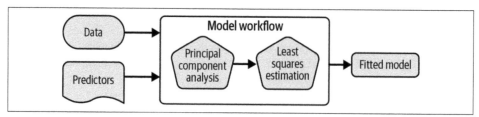

Figure 7-2. Correct mental model of where model estimation occurs in the data analysis process.

In this way, the PCA preprocessing is considered part of the modeling process.

Workflow Basics

The workflows package allows the user to bind modeling and preprocessing objects together. Let's start again with the Ames data and a simple linear model:

```
library(tidymodels)  # Includes the workflows package
tidymodels_prefer()

lm_model <-
  linear_reg() %>%
  set_engine("lm")
```

A workflow always requires a parsnip model object:

```
lm_wflow <-
  workflow() %>%
  add_model(lm_model)

lm_wflow
#> == Workflow =====================================
#> Preprocessor: None
#> Model: linear_reg()
#>
#> -- Model ----------------------------------------
#> Linear Regression Model Specification (regression)
#>
#> Computational engine: lm
```

Notice that we have not yet specified how this workflow should preprocess the data: `Preprocessor: None`.

If our model is very simple, a standard R formula can be used as a preprocessor:

```
lm_wflow <-
  lm_wflow %>%
  add_formula(Sale_Price ~ Longitude + Latitude)

lm_wflow
#> == Workflow =====================================
#> Preprocessor: Formula
#> Model: linear_reg()
#>
```

```
#> — Preprocessor ————————————————————————
#> Sale_Price ~ Longitude + Latitude
#>
#> — Model ————————————————————————————————
#> Linear Regression Model Specification (regression)
#>
#> Computational engine: lm
```

Workflows have a `fit()` method that can be used to create the model. Using the objects created in the summary at the end of Chapter 6 we see the following:

```
lm_fit <- fit(lm_wflow, ames_train)
lm_fit
#> ══ Workflow [trained] ══════════════════════════
#> Preprocessor: Formula
#> Model: linear_reg()
#>
#> — Preprocessor ————————————————————————
#> Sale_Price ~ Longitude + Latitude
#>
#> — Model ————————————————————————————————
#>
#> Call:
#> stats::lm(formula = ..y ~ ., data = data)
#>
#> Coefficients:
#> (Intercept)    Longitude     Latitude
#>     -302.97        -2.07         2.71
```

We can also `predict()` on the fitted workflow:

```
predict(lm_fit, ames_test %>% slice(1:3))
#> # A tibble: 3 × 1
#>   .pred
#>   <dbl>
#> 1  5.22
#> 2  5.21
#> 3  5.28
```

The `predict()` method follows all of the same rules and naming conventions that we described for the parsnip package in Chapter 6.

Both the model and preprocessor can be removed or updated:

```
lm_fit %>% update_formula(Sale_Price ~ Longitude)
#> ══ Workflow ════════════════════════════════════
#> Preprocessor: Formula
#> Model: linear_reg()
#>
#> — Preprocessor ————————————————————————
#> Sale_Price ~ Longitude
#>
#> — Model ————————————————————————————————
#> Linear Regression Model Specification (regression)
#>
#> Computational engine: lm
```

Note that, in this new object, the output shows that the previous fitted model was removed since the new formula is inconsistent with the previous model fit.

Adding Raw Variables to the workflow()

There is another interface for passing data to the model, the add_variables()
function, which uses a dplyr-like syntax for choosing variables. The function has
two primary arguments: outcomes and predictors. These use a selection approach
similar to the tidyselect backend of tidyverse packages to capture multiple selectors
using c():

```
lm_wflow <-
  lm_wflow %>%
  remove_formula() %>%
  add_variables(outcome = Sale_Price, predictors = c(Longitude, Latitude))
lm_wflow
#> ══ Workflow ═══════════════════════════════════════════
#> Preprocessor: Variables
#> Model: linear_reg()
#>
#> ── Preprocessor ───────────────────────────────────────
#> Outcomes: Sale_Price
#> Predictors: c(Longitude, Latitude)
#>
#> ── Model ──────────────────────────────────────────────
#> Linear Regression Model Specification (regression)
#>
#> Computational engine: lm
```

The predictors could also have been specified using a more general selector, such as:

```
predictors = c(ends_with("tude"))
```

One nicety is that any outcome columns accidentally specified in the predictors
argument will be quietly removed. This facilitates the use of:

```
predictors = everything()
```

When the model is fit, the specification assembles these data, unaltered, into a data
frame and passes it to the underlying function:

```
fit(lm_wflow, ames_train)
#> ══ Workflow [trained] ═════════════════════════════════
#> Preprocessor: Variables
#> Model: linear_reg()
#>
#> ── Preprocessor ───────────────────────────────────────
#> Outcomes: Sale_Price
#> Predictors: c(Longitude, Latitude)
#>
#> ── Model ──────────────────────────────────────────────
#>
#> Call:
#> stats::lm(formula = ..y ~ ., data = data)
#>
#> Coefficients:
#> (Intercept)    Longitude     Latitude
#>     -302.97        -2.07         2.71
```

If you would like the underlying modeling method to do what it would normally do with the data, add_variables() can be a helpful interface. As we will see in an upcoming section in this chapter, it also facilitates more complex modeling specifications. However, as we mention in the next section, models such as glmnet and xgboost expect the user to make indicator variables from factor predictors. In these cases, a recipe or formula interface will typically be a better choice.

In the next chapter, we will look at a more powerful preprocessor (called a *recipe*) that can also be added to a workflow.

How Does a workflow() Use the Formula?

Recall from Chapter 3 that the formula method in R has multiple purposes (we will discuss this further in Chapter 8). One of these is to properly encode the original data into an analysis-ready format. This can involve executing inline transformations (e.g., log(x)), creating dummy variable columns, creating interactions or other column expansions, and so on. However, many statistical methods require different types of encodings:

- Most packages for tree-based models use the formula interface but *do not* encode the categorical predictors as dummy variables.

- Packages can use special inline functions that tell the model function how to treat the predictor in the analysis. For example, in survival analysis models, a formula term such as strata(site) would indicate that the column site is a stratification variable. This means it should not be treated as a regular predictor and does not have a corresponding location parameter estimate in the model.

- A few R packages have extended the formula in ways that base R functions cannot parse or execute. In multilevel models (e.g., mixed models or hierarchical Bayesian models), a model term such as (week | subject) indicates that the column week is a random effect that has different slope parameter estimates for each value of the subject column.

A workflow is a general purpose interface. When add_formula() is used, how should the workflow preprocess the data? Since the preprocessing is model dependent, workflows attempts to emulate what the underlying model would do whenever possible. If it is not possible, the formula processing should not do anything to the columns used in the formula. Let's look at this in more detail.

Tree-Based Models

When we fit a tree to the data, the parsnip package understands what the modeling function would do. For example, if a random forest model is fit using the ranger or randomForest packages, the workflow knows predictors columns that are factors should be left as is.

As a counterexample, a boosted tree created with the xgboost package requires the user to create dummy variables from factor predictors (since `xgboost::xgb.train()` will not). This requirement is embedded into the model specification object, and a workflow using xgboost will create the indicator columns for this engine. Also note that a different engine for boosted trees, C5.0, does not require dummy variables so none are made by the workflow.

This determination is made for each model and engine combination.

Special Formulas and Inline Functions

A number of multilevel models have been standardized as a formula specification devised in the lme4 package. For example, to fit a regression model that has random effects for subjects, we would use the following formula:

```
library(lme4)
lmer(distance ~ Sex + (age | Subject), data = Orthodont)
```

The effect is that each subject will have an estimated intercept and slope parameter for age.

The problem is that standard R methods can't properly process this formula:

```
model.matrix(distance ~ Sex + (age | Subject), data = Orthodont)
#> Warning in Ops.ordered(age, Subject): '|' is not meaningful for ordered factors
#>     (Intercept) SexFemale age | SubjectTRUE
#> attr(,"assign")
#> [1] 0 1 2
#> attr(,"contrasts")
#> attr(,"contrasts")$Sex
#> [1] "contr.treatment"
#>
#> attr(,"contrasts")$`age | Subject`
#> [1] "contr.treatment"
```

The result is a zero row data frame.

 The issue is that the special formula has to be processed by the underlying package code, not the standard model.matrix() approach.

Even if this formula could be used with model.matrix(), this would still present a problem since the formula also specifies the statistical attributes of the model.

The solution in workflows is an optional supplementary model formula that can be passed to add_model(). The add_variables() specification provides the bare column names, and then the actual formula given to the model is set within add_model():

```
library(multilevelmod)

multilevel_spec <- linear_reg() %>% set_engine("lmer")

multilevel_workflow <-
  workflow() %>%
  # Pass the data along as-is:
  add_variables(outcome = distance, predictors = c(Sex, age, Subject)) %>%
  add_model(multilevel_spec,
            # This formula is given to the model
            formula = distance ~ Sex + (age | Subject))

multilevel_fit <- fit(multilevel_workflow, data = Orthodont)
multilevel_fit
#> ══ Workflow [trained] ═══════════════════════════════
#> Preprocessor: Variables
#> Model: linear_reg()
#>
#> ── Preprocessor ─────────────────────────────────────
#> Outcomes: distance
#> Predictors: c(Sex, age, Subject)
#>
#> ── Model ────────────────────────────────────────────
#> Linear mixed model fit by REML ['lmerMod']
#> Formula: distance ~ Sex + (age | Subject)
#>    Data: data
#> REML criterion at convergence: 471.2
#> Random effects:
#>  Groups   Name        Std.Dev. Corr
#>  Subject  (Intercept) 7.391
#>           age         0.694    -0.97
#>  Residual             1.310
#> Number of obs: 108, groups:  Subject, 27
#> Fixed Effects:
#> (Intercept)    SexFemale
#>       24.52        -2.15
```

We can even use the previously mentioned strata() function from the survival package for survival analysis:

```
library(censored)

parametric_spec <- survival_reg()

parametric_workflow <-
  workflow() %>%
  add_variables(outcome = c(fustat, futime), predictors = c(age, rx)) %>%
  add_model(parametric_spec,
            formula = Surv(futime, fustat) ~ age + strata(rx))

parametric_fit <- fit(parametric_workflow, data = ovarian)
```

```
parametric_fit
#> ══ Workflow [trained] ═══════════════════════════════════
#> Preprocessor: Variables
#> Model: survival_reg()
#>
#> ── Preprocessor ─────────────────────────────────────────
#> Outcomes: c(fustat, futime)
#> Predictors: c(age, rx)
#>
#> ── Model ────────────────────────────────────────────────
#> Call:
#> survival::survreg(formula = Surv(futime, fustat) ~ age + strata(rx),
#>     data = data, model = TRUE)
#>
#> Coefficients:
#> (Intercept)          age
#>     12.8734      -0.1034
#>
#> Scale:
#>   rx=1    rx=2
#> 0.7696 0.4704
#>
#> Loglik(model)= -89.4   Loglik(intercept only)= -97.1
#>   Chisq= 15.36 on 1 degrees of freedom, p= 9e-05
#> n= 26
```

Notice how in both of these calls the model-specific formula was used.

Creating Multiple Workflows at Once

In some situations the data require numerous attempts to find an appropriate model. For example:

- For predictive models, it is advisable to evaluate a variety of different model types. This requires the user to create multiple model specifications.

- Sequential testing of models typically starts with an expanded set of predictors. This "full model" is compared to a sequence of the same model that removes each predictor in turn. Using basic hypothesis testing methods or empirical validation, the effect of each predictor can be isolated and assessed.

In these situations, as well as others, it can become tedious or onerous to create a lot of workflows from different sets of preprocessors and/or model specifications. To address this problem, the workflowset package creates combinations of workflow components. A list of preprocessors (e.g., formulas, dplyr selectors, or feature engineering recipe objects discussed in Chapter 8) can be combined with a list of model specifications, resulting in a set of workflows.

As an example, let's say that we want to focus on the different ways that house location is represented in the Ames data. We can create a set of formulas that capture these predictors:

```
location <- list(
  longitude = Sale_Price ~ Longitude,
  latitude = Sale_Price ~ Latitude,
  coords = Sale_Price ~ Longitude + Latitude,
  neighborhood = Sale_Price ~ Neighborhood
)
```

These representations can be crossed with one or more models using the
workflow_set() function. We'll just use the previous linear model specification to
demonstrate:

```
library(workflowsets)
location_models <- workflow_set(preproc = location, models = list(lm = lm_model))
location_models
#> # A workflow set/tibble: 4 × 4
#>   wflow_id        info             option    result
#>   <chr>           <list>           <list>    <list>
#> 1 longitude_lm    <tibble [1 × 4]> <opts[0]> <list [0]>
#> 2 latitude_lm     <tibble [1 × 4]> <opts[0]> <list [0]>
#> 3 coords_lm       <tibble [1 × 4]> <opts[0]> <list [0]>
#> 4 neighborhood_lm <tibble [1 × 4]> <opts[0]> <list [0]>
location_models$info[[1]]
#> # A tibble: 1 × 4
#>   workflow    preproc model      comment
#>   <list>      <chr>   <chr>      <chr>
#> 1 <workflow> formula linear_reg ""
extract_workflow(location_models, id = "coords_lm")
#> ══ Workflow ═══════════════════════════════════════════
#> Preprocessor: Formula
#> Model: linear_reg()
#>
#> ── Preprocessor ───────────────────────────────────────
#> Sale_Price ~ Longitude + Latitude
#>
#> ── Model ──────────────────────────────────────────────
#> Linear Regression Model Specification (regression)
#>
#> Computational engine: lm
```

Workflow sets are mostly designed to work with resampling, which is discussed in
Chapter 10. The columns option and result must be populated with specific types
of objects that result from resampling. We will demonstrate this in more detail in
Chapters 11 and 15.

In the meantime, let's create model fits for each formula and save them in a new
column called fit. We'll use basic dplyr and purrr operations:

```
location_models <-
  location_models %>%
  mutate(fit = map(info, ~ fit(.x$workflow[[1]], ames_train)))
location_models
#> # A workflow set/tibble: 4 × 5
#>   wflow_id        info             option    result     fit
#>   <chr>           <list>           <list>    <list>     <list>
#> 1 longitude_lm    <tibble [1 × 4]> <opts[0]> <list [0]> <workflow>
#> 2 latitude_lm     <tibble [1 × 4]> <opts[0]> <list [0]> <workflow>
#> 3 coords_lm       <tibble [1 × 4]> <opts[0]> <list [0]> <workflow>
```

```
#> 4 neighborhood_lm <tibble [1 x 4]> <opts[0]> <list [0]> <workflow>
location_models$fit[[1]]
#> ══ Workflow [trained] ══════════════════════════════════
#> Preprocessor: Formula
#> Model: linear_reg()
#>
#> ── Preprocessor ────────────────────────────────────────
#> Sale_Price ~ Longitude
#>
#> ── Model ───────────────────────────────────────────────
#>
#> Call:
#> stats::lm(formula = ..y ~ ., data = data)
#>
#> Coefficients:
#> (Intercept)     Longitude
#>     -184.40         -2.02
```

We use a purrr function here to map through our models, but there is an easier, better approach to fit workflow sets that will be introduced in Chapter 11.

In general, there's a lot more to workflow sets! While we've covered the basics here, the nuances and advantages of workflow sets won't be illustrated until Chapter 15.

Evaluating the Test Set

Let's say that we've concluded our model development and have settled on a final model. There is a convenience function called `last_fit()` that will *fit* the model to the entire training set and *evaluate* it with the testing set.

Using `lm_wflow` as an example, we can pass the model and the initial training/testing split to the function:

```
final_lm_res <- last_fit(lm_wflow, ames_split)
final_lm_res
#> # Resampling results
#> # Manual resampling
#> # A tibble: 1 × 6
#>   splits            id              .metrics .notes  .predictions .workflow
#>   <list>            <chr>           <list>   <list>  <list>       <list>
#> 1 <split [2342/588]> train/test split <tibble> <tibble> <tibble>     <workflow>
```

Notice that `last_fit()` takes a data split as an input, not a data frame. This function uses the split to generate the training and test sets for the final fitting and evaluation.

The .workflow column contains the fitted workflow and can be pulled out of the results using:

```
fitted_lm_wflow <- extract_workflow(final_lm_res)
```

Similarly, collect_metrics() and collect_predictions() provide access to the performance metrics and predictions, respectively:

```
collect_metrics(final_lm_res)
collect_predictions(final_lm_res) %>% slice(1:5)
```

We'll see more about last_fit() in action and how to use it again in Chapter 16.

Chapter Summary

In this chapter, you learned that the modeling process encompasses more than just estimating the parameters of an algorithm that connects predictors to an outcome. This process also includes preprocessing steps and operations taken after a model is fit. We introduced a concept called a *model workflow* that can capture the important components of the modeling process. Multiple workflows can also be created inside of a *workflow set*. The last_fit() function is convenient for fitting a final model to the training set and evaluating with the test set.

For the Ames data, the related code that we'll see used again is:

```
library(tidymodels)
data(ames)

ames <- mutate(ames, Sale_Price = log10(Sale_Price))

set.seed(502)
ames_split <- initial_split(ames, prop = 0.80, strata = Sale_Price)
ames_train <- training(ames_split)
ames_test  <- testing(ames_split)

lm_model <- linear_reg() %>% set_engine("lm")

lm_wflow <-
  workflow() %>%
  add_model(lm_model) %>%
  add_variables(outcome = Sale_Price, predictors = c(Longitude, Latitude))

lm_fit <- fit(lm_wflow, ames_train)
```

## CHAPTER 8

Feature Engineering with Recipes

Feature engineering entails reformatting predictor values to make them easier for a model to use effectively. This includes transformations and encodings of the data to best represent their important characteristics. Imagine that you have two predictors in a data set that can be more effectively represented in your model as a ratio; creating a new predictor from the ratio of the original two is a simple example of feature engineering.

Take the location of a house in Ames as a more involved example. There are a variety of ways that this spatial information can be exposed to a model, including neighborhood (a qualitative measure), longitude/latitude, distance to the nearest school, and so on. When choosing how to encode these data in modeling, we might choose an option we believe is most associated with the outcome. The original format of the data, for example numeric (e.g., distance) versus categorical (e.g., neighborhood), is also a driving factor in feature engineering choices.

Other examples of preprocessing to build better features for modeling include:

- Correlation between predictors can be reduced via feature extraction or the removal of some predictors.

- When some predictors have missing values, they can be imputed using a sub-model.

- Models that use variance-type measures may benefit from coercing the distribution of some skewed predictors to be symmetric by estimating a transformation.

Feature engineering and data preprocessing can also involve reformatting that may be required by the model. Some models use geometric distance metrics and, consequently, numeric predictors should be centered and scaled so that they are all in the same units. Otherwise, the distance values would be biased by the scale of each column.

 Different models have different preprocessing requirements and some, such as tree-based models, require very little preprocessing at all. The Appendix contains a small table of recommended preprocessing techniques for different models.

In this chapter, we introduce the recipes package (*https://oreil.ly/b34bX*), which you can use to combine different feature engineering and preprocessing tasks into a single object and then apply these transformations to different data sets. The recipes package is, like parsnip for models, one of the core tidymodels packages.

This chapter uses the Ames housing data and the R objects created in the book so far, as summarized at the end of Chapter 7.

A Simple recipe() for the Ames Housing Data

In this section, we will focus on a small subset of the predictors available in the Ames housing data:

- The neighborhood (qualitative, with 29 neighborhoods in the training set)
- The gross above-grade living area (continuous, named Gr_Liv_Area)
- The year built (Year_Built)
- The type of building (Bldg_Type with values OneFam ($n = 1,936$), TwoFmCon ($n = 50$), Duplex ($n = 88$), Twnhs ($n = 77$), and TwnhsE ($n = 191$))

Suppose that an initial ordinary linear regression model were fit to these data. Recalling that, in Chapter 4, the sale prices were prelogged, a standard call to lm() might look like:

```
lm(Sale_Price ~ Neighborhood + log10(Gr_Liv_Area) + Year_Built + Bldg_Type, data = ames)
```

When this function is executed, the data are converted from a data frame to a numeric *design matrix* (also called a *model matrix*) and then the least squares method is used to estimate parameters. In Chapter 3 we listed the multiple purposes of the R model formula; let's focus only on the data manipulation aspects for now. What the preceding formula does can be decomposed into a series of steps:

1. Sale price is defined as the outcome while neighborhood, gross living area, the year built, and building type variables are all defined as predictors.

2. A log transformation is applied to the gross living area predictor.

3. The neighborhood and building type columns are converted from a nonnumeric format to a numeric format (since least squares requires numeric predictors).

As mentioned in Chapter 3, the formula method will apply these data manipulations to any data, including new data, that are passed to the `predict()` function.

A recipe is also an object that defines a series of steps for data processing. Unlike the formula method inside a modeling function, the recipe defines the steps via `step_*()` functions without immediately executing them; it is only a specification of what should be done. Here is a recipe equivalent to the preceding formula that builds on the code summary at the end of Chapter 5:

```
library(tidymodels) # Includes the recipes package
tidymodels_prefer()

simple_ames <-
  recipe(Sale_Price ~ Neighborhood + Gr_Liv_Area + Year_Built + Bldg_Type,
         data = ames_train) %>%
  step_log(Gr_Liv_Area, base = 10) %>%
  step_dummy(all_nominal_predictors())
simple_ames
#> Recipe
#>
#> Inputs:
#>
#>       role #variables
#>    outcome          1
#>  predictor          4
#>
#> Operations:
#>
#> Log transformation on Gr_Liv_Area
#> Dummy variables from all_nominal_predictors()
```

Let's break this down:

1. The call to `recipe()` with a formula tells the recipe the *roles* of the "ingredients" or variables (e.g., predictor, outcome). It only uses the data `ames_train` to determine the data types for the columns.

2. `step_log()` declares that `Gr_Liv_Area` should be log transformed.

3. `step_dummy()` specifies which variables should be converted from a qualitative format to a quantitative format, in this case, using dummy or indicator variables. An indicator or dummy variable is a binary numeric variable (a column of ones and zeroes) that encodes qualitative information; we will dig deeper into these kinds of variables later in this chapter.

The function `all_nominal_predictors()` captures the names of any predictor columns that are currently factor or character (i.e., nominal) in nature. This is a dplyr-like selector function similar to `starts_with()` or `matches()` but that can only be used inside of a recipe.

 Other selectors specific to the recipes package are: `all_numeric_predictors()`, `all_numeric()`, `all_predictors()`, and `all_outcomes()`. As with dplyr, one or more unquoted expressions, separated by commas, can be used to select which columns are affected by each step.

What is the advantage to using a recipe, over a formula or raw predictors? There are a few, including:

- These computations can be recycled across models since they are not tightly coupled to the modeling function.
- A recipe enables a broader set of data processing choices than formulas can offer.
- The syntax can be very compact. For example, `all_nominal_predictors()` can be used to capture many variables for specific types of processing while a formula would require each to be explicitly listed.
- All data processing can be captured in a single R object instead of in scripts that are repeated, or even spread across different files.

Using Recipes

As we discussed in Chapter 7, preprocessing choices and feature engineering should typically be considered part of a modeling workflow, not a separate task. The workflows package contains high-level functions to handle different types of preprocessors. Our previous workflow (`lm_wflow`) used a simple set of dplyr selectors. To improve on that approach with more complex feature engineering, let's use the `simple_ames` recipe to preprocess data for modeling.

This object can be attached to the workflow:

```
lm_wflow %>%
  add_recipe(simple_ames)
#> Error in `add_recipe()`:
#> ! A recipe cannot be added when variables already exist.
```

That did not work! We can have only one preprocessing method at a time, so we need to remove the existing preprocessor before adding the recipe.

```
lm_wflow <-
  lm_wflow %>%
  remove_variables() %>%
  add_recipe(simple_ames)
lm_wflow
#> ══ Workflow ═══════════════════════════════════
#> Preprocessor: Recipe
```

```
#> Model: linear_reg()
#>
#> — Preprocessor ─────────────────────────────────
#> 2 Recipe Steps
#>
#> • step_log()
#> • step_dummy()
#>
#> — Model ────────────────────────────────────────
#> Linear Regression Model Specification (regression)
#>
#> Computational engine: lm
```

Let's estimate both the recipe and model using a simple call to `fit()`:

```
lm_fit <- fit(lm_wflow, ames_train)
```

The `predict()` method applies the same preprocessing that was used on the training set to the new data before passing them along to the model's `predict()` method:

```
predict(lm_fit, ames_test %>% slice(1:3))
#> # A tibble: 3 × 1
#>    .pred
#>    <dbl>
#> 1   5.08
#> 2   5.32
#> 3   5.28
```

If we need the bare model object or recipe, there are `extract_*` functions that can retrieve them:

```
# Get the recipe after it has been estimated:
lm_fit %>%
  extract_recipe(estimated = TRUE)
#> Recipe
#>
#> Inputs:
#>
#>       role #variables
#>    outcome          1
#>  predictor          4
#>
#> Training data contained 2342 data points and no missing data.
#>
#> Operations:
#>
#> Log transformation on Gr_Liv_Area [trained]
#> Dummy variables from Neighborhood, Bldg_Type [trained]

# To tidy the model fit:
lm_fit %>%
  # This returns the parsnip object:
  extract_fit_parsnip() %>%
  # Now tidy the linear model object:
  tidy() %>%
  slice(1:5)
#> # A tibble: 5 × 5
#>    term                estimate std.error statistic  p.value
#>    <chr>                  <dbl>     <dbl>     <dbl>    <dbl>
```

```
#> 1 (Intercept)                      -0.669   0.231      -2.90 3.80e-  3
#> 2 Gr_Liv_Area                       0.620   0.0143      43.2 2.63e-299
#> 3 Year_Built                        0.00200 0.000117    17.1 6.16e- 62
#> 4 Neighborhood_College_Creek  0.0178   0.00819      2.17 3.02e-  2
#> 5 Neighborhood_Old_Town            -0.0330   0.00838     -3.93 8.66e-  5
```

Tools for using (and debugging) recipes outside of workflow objects are described in Chapter 16.

How Data Are Used by the recipe()

Data are passed to recipes at different stages.

First, when calling recipe(..., data), the data set is used to determine the data types of each column so that selectors such as all_numeric() or all_numeric _predictors() can be used.

Second, when preparing the data using fit(workflow, data), the training data are used for all estimation operations including a recipe that may be part of the workflow, from determining factor levels to computing PCA components and everything in between.

All preprocessing and feature engineering steps use *only* the training data. Otherwise, information leakage can negatively impact the model's performance when used with new data.

Finally, when using predict(workflow, new_data), no model or preprocessor parameters like those from recipes are reestimated using the values in new_data. Take centering and scaling using step_normalize() as an example. Using this step, the means and standard deviations from the appropriate columns are determined from the training set; new samples at prediction time are standardized using these values from training when predict() is invoked.

Examples of Steps

Before proceeding, let's take an extended tour of the capabilities of recipes and explore some of the most important step_*() functions. These recipe step functions each specify a specific possible step in a feature engineering process, and different recipe steps can have different effects on columns of data.

Encoding Qualitative Data in a Numeric Format

One of the most common feature engineering tasks is transforming nominal or qualitative data (factors or characters) so that they can be encoded or represented numerically. Sometimes we can alter the factor levels of a qualitative column in helpful ways prior to such a transformation. For example, step_unknown() can be used to change missing values to a dedicated factor level. Similarly, if we anticipate that a new factor level may be encountered in future data, step_novel() can allot a new level for this purpose.

Additionally, step_other() can be used to analyze the frequencies of the factor levels in the training set and convert infrequently occurring values to a catch-all level of "other," with a threshold that can be specified. A good example is the Neighborhood predictor in our data, shown in Figure 8-1.

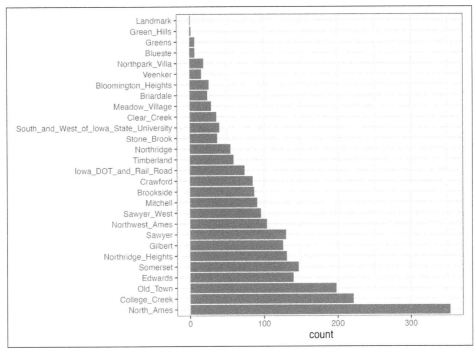

Figure 8-1. Frequencies of neighborhoods in the Ames training set.

Here we see that two neighborhoods have less than five properties in the training data (Landmark and Green Hills); in this case, no houses at all in the Landmark neighborhood were included in the training set. For some models, it may be problematic to have dummy variables with a single nonzero entry in the column. At a minimum, it is highly improbable that these features would be important to a model. If we add step_other(Neighborhood, threshold = 0.01) to our recipe, the bottom 1% of the neighborhoods will be lumped into a new level called "other." In this training set, this will catch seven neighborhoods.

For the Ames data, we can amend the recipe to use:

```
simple_ames <-
  recipe(Sale_Price ~ Neighborhood + Gr_Liv_Area + Year_Built + Bldg_Type,
         data = ames_train) %>%
  step_log(Gr_Liv_Area, base = 10) %>%
  step_other(Neighborhood, threshold = 0.01) %>%
  step_dummy(all_nominal_predictors())
```

Many, but not all, underlying model calculations require predictor values to be encoded as numbers. Notable exceptions include tree-based models, rule-based models, and naive Bayes models.

The most common method for converting a factor predictor to a numeric format is to create dummy or indicator variables. Let's take the predictor in the Ames data for the building type, which is a factor variable with five levels (see Table 8-1). For dummy variables, the single Bldg_Type column would be replaced with four numeric columns whose values are either zero or one. These binary variables represent specific factor-level values. In R, the convention is to exclude a column for the first factor level (OneFam, in this case). The Bldg_Type column would be replaced with a column called TwoFmCon that is one when the row has that value and zero otherwise. Three other columns are similarly created:

Table 8-1. Illustration of binary encodings (i.e., dummy variables) for a qualitative predictor

Raw data	TwoFmCon	Duplex	Twnhs	TwnhsE
OneFam	0	0	0	0
TwoFmCon	1	0	0	0
Duplex	0	1	0	0
Twnhs	0	0	1	0
TwnhsE	0	0	0	1

Why not all five? The most basic reason is simplicity; if you know the value for these four columns, you can determine the last value because these are mutually exclusive categories. More technically, the classical justification is that a number of models, including ordinary linear regression, have numerical issues when there are linear dependencies between columns. If all five building type indicator columns are included, they would add up to the intercept column (if there is one). This would cause an issue, or perhaps an outright error, in the underlying matrix algebra.

The full set of encodings can be used for some models. This is traditionally called *one-hot encoding* and can be achieved using the one_hot argument of step_dummy().

One helpful feature of step_dummy() is that there is more control over how the resulting dummy variables are named. In base R, dummy variable names mash the variable name with the level, resulting in names like NeighborhoodVeenker. Recipes, by default, use an underscore as the separator between the name and level (e.g., Neighborhood_Veenker), and there is an option to use custom formatting for the names. The default naming convention in recipes makes it easier to capture those new columns in future steps using a selector, such as starts_with("Neighbor hood_").

Traditional dummy variables require that all of the possible categories be known to create a full set of numeric features. There are other methods for doing this transformation to a numeric format. *Feature hashing* methods only consider the value of the category to assign it to a predefined pool of dummy variables. *Effect* or *likelihood encodings* replace the original data with a single numeric column that measures the *effect* of those data. Both feature hashing and effect encoding can seamlessly handle situations where a novel factor level is encountered in the data. Chapter 17 explores these and other methods for encoding categorical data, beyond straightforward dummy or indicator variables.

 Different recipe steps behave differently when applied to variables in the data. For example, step_log() modifies a column in place without changing the name. Other steps, such as step_dummy(), eliminate the original data column and replace it with one or more columns with different names. The effect of a recipe step depends on the type of feature engineering transformation being done.

Interaction Terms

Interaction effects involve two or more predictors. Such an effect occurs when one predictor has an effect on the outcome that is contingent on one or more other predictors. For example, if you were trying to predict how much traffic there will be during your commute, two potential predictors could be the specific time of day you commute and the weather. However, the relationship between the amount of traffic and bad weather is different for different times of day. In this case, you could add an interaction term between the two predictors to the model along with the original two predictors (which are called the main effects). Numerically, an interaction term between predictors is encoded as their product. Interactions are defined in terms of their effect on the outcome and can be combinations of different types of data (e.g., numeric, categorical, etc.). Chapter 7 of Kuhn and Johnson (2020) (*https://oreil.ly/WCpCP*) discusses interactions and how to detect them in greater detail.

After exploring the Ames training set, we might find that the regression slopes for the gross living area differ for different building types, as shown in Figure 8-2:

```
ggplot(ames_train, aes(x = Gr_Liv_Area, y = 10^Sale_Price)) +
  geom_point(alpha = .2) +
  facet_wrap(~ Bldg_Type) +
  geom_smooth(method = lm, formula = y ~ x, se = FALSE, color = "lightblue") +
  scale_x_log10() +
  scale_y_log10() +
  labs(x = "Gross Living Area", y = "Sale Price (USD)")
```

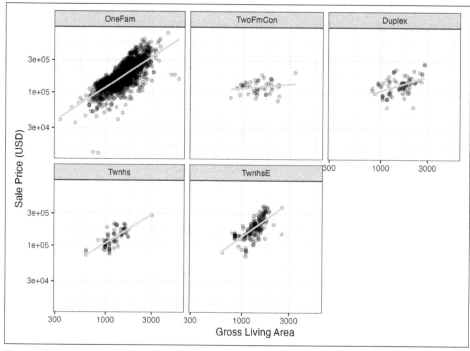

Figure 8-2. Gross living area (in log-10 units) versus sale price (also in log-10 units) for five different building types.

How are interactions specified in a recipe? A base R formula would take an interaction using a :, so we would use:

```
Sale_Price ~ Neighborhood + log10(Gr_Liv_Area) + Bldg_Type +
  log10(Gr_Liv_Area):Bldg_Type
# or
Sale_Price ~ Neighborhood + log10(Gr_Liv_Area) * Bldg_Type
```

where * expands those columns to the main effects and interaction term. Again, the formula method does many things simultaneously and understands that a factor variable (such as Bldg_Type) should be expanded into dummy variables first and that the interaction should involve all of the resulting binary columns.

Recipes are more explicit and sequential, and they give you more control. With the current recipe, step_dummy() has already created dummy variables. How would we combine these for an interaction? The additional step would look like step `_interact(~ interaction terms)` where the terms on the right-hand side of the tilde are the interactions. These can include selectors, so it would be appropriate to use:

```
simple_ames <-
  recipe(Sale_Price ~ Neighborhood + Gr_Liv_Area + Year_Built + Bldg_Type,
         data = ames_train) %>%
  step_log(Gr_Liv_Area, base = 10) %>%
  step_other(Neighborhood, threshold = 0.01) %>%
  step_dummy(all_nominal_predictors()) %>%
  # Gr_Liv_Area is on the log scale from a previous step
  step_interact( ~ Gr_Liv_Area:starts_with("Bldg_Type_") )
```

Additional interactions can be specified in this formula by separating them by +. Also note that the recipe will only use interactions between different variables; if the formula uses `var_1:var_1`, this term will be ignored.

Suppose that, in a recipe, we had not yet made dummy variables for building types. It would be inappropriate to include a factor column in this step, such as:

```
step_interact( ~ Gr_Liv_Area:Bldg_Type )
```

This is telling the underlying (base R) code used by `step_interact()` to make dummy variables and then form the interactions. In fact, if this occurs, a warning states that this might generate unexpected results.

This behavior gives you more control, but it is different from R's standard model formula.

As with naming dummy variables, recipes provides more coherent names for interaction terms. In this case, the interaction is named `Gr_Liv_Area_x_Bldg_Type_Duplex` instead of `Gr_Liv_Area:Bldg_TypeDuplex` (which is not a valid column name for a data frame).

Remember that order matters. The gross living area is log transformed prior to the interaction term. Subsequent interactions with this variable will also use the log scale.

Spline Functions

When a predictor has a nonlinear relationship with the outcome, some types of predictive models can adaptively approximate this relationship during training. However, simpler is usually better, and it is not uncommon to try to use a simple model, such as a linear fit, and add in specific nonlinear features for predictors that may need them, such as longitude and latitude for the Ames housing data. One common method for doing this is to use *spline* functions to represent the data. Splines replace the existing numeric predictor with a set of columns that allow a model to emulate a flexible, nonlinear relationship. As more spline terms are added to the data, the capacity to nonlinearly represent the relationship increases. Unfortunately, it may also increase the likelihood of picking up on data trends that occur by chance (i.e., overfitting).

If you have ever used `geom_smooth()` within a `ggplot`, you have probably used a spline representation of the data. For example, each panel in Figure 8-3 uses a different number of smooth splines for the latitude predictor:

```
library(patchwork)
library(splines)

plot_smoother <- function(deg_free) {
  ggplot(ames_train, aes(x = Latitude, y = 10^Sale_Price)) +
    geom_point(alpha = .2) +
    scale_y_log10() +
    geom_smooth(
      method = lm,
      formula = y ~ ns(x, df = deg_free),
      color = "lightblue",
      se = FALSE
    ) +
    labs(title = paste(deg_free, "Spline Terms"),
         y = "Sale Price (USD)")
}

( plot_smoother(2) + plot_smoother(5) ) / ( plot_smoother(20) + plot_smoother(100) )
```

The `ns()` function in the splines package generates feature columns using functions called *natural splines*.

Some panels in Figure 8-3 fit poorly; 2 terms *underfit* the data while 100 terms *overfit*. The panels with 5 and 20 terms seem like reasonably smooth fits that catch the main patterns of the data. This indicates that the proper amount of "nonlinearness" matters. The number of spline terms could then be considered a *tuning parameter* for this model. These types of parameters are explored in Chapter 12.

In recipes, multiple steps can create these types of terms. To add a natural spline representation for this predictor:

```
recipe(Sale_Price ~ Neighborhood + Gr_Liv_Area + Year_Built + Bldg_Type + Latitude,
       data = ames_train) %>%
  step_log(Gr_Liv_Area, base = 10) %>%
  step_other(Neighborhood, threshold = 0.01) %>%
  step_dummy(all_nominal_predictors()) %>%
  step_interact( ~ Gr_Liv_Area:starts_with("Bldg_Type_") ) %>%
  step_ns(Latitude, deg_free = 20)
```

The user would need to determine if both neighborhood and latitude should be in the model since they both represent the same underlying data in different ways.

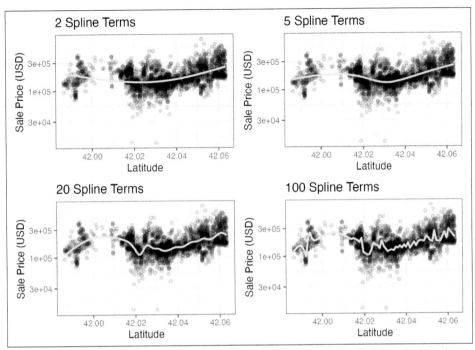

Figure 8-3. Sale price versus latitude, with trend lines using natural splines with different degrees of freedom.

Feature Extraction

Another common method for representing multiple features at once is called *feature extraction*. Most of these techniques create new features from the predictors that capture the information in the broader set as a whole. For example, principal component analysis (PCA) tries to extract as much of the original information in the predictor set as possible using a smaller number of features. PCA is a linear extraction method, meaning that each new feature is a linear combination of the original predictors. One nice aspect of PCA is that each of the new features, called the principal components or PCA scores, are uncorrelated with one another. Because of this, PCA can be very effective at reducing the correlation between predictors. Note that PCA is only aware of the predictors; the new PCA features might not be associated with the outcome.

In the Ames data, several predictors measure size of the property, such as the total basement size (`Total_Bsmt_SF`), size of the first floor (`First_Flr_SF`), the gross living area (`Gr_Liv_Area`), and so on. PCA might be an option to represent these potentially redundant variables as a smaller feature set. Apart from the gross living area, these predictors have the suffix `SF` in their names (for square feet), so a recipe step for PCA might look like:

```
# Use a regular expression to capture house size predictors:
step_pca(matches("(SF$)|(Gr_Liv)"))
```

Note that all of these columns are measured in square feet. PCA assumes that all of the predictors are on the same scale. That's true in this case, but often this step can be preceded by `step_normalize()`, which will center and scale each column.

There are existing recipe steps for other extraction methods, such as independent component analysis (ICA), nonnegative matrix factorization (NNMF), multidimensional scaling (MDS), uniform manifold approximation and projection (UMAP), and others.

Row Sampling Steps

Recipe steps can affect the rows of a data set as well. For example, *subsampling* techniques for class imbalances change the class proportions in the data being given to the model; these techniques often don't improve overall performance but can generate better behaved distributions of the predicted class probabilities. These are approaches to try when subsampling your data with class imbalance:

Downsampling
Downsampling the data keeps the minority class and takes a random sample of the majority class so that class frequencies are balanced.

Upsampling
Upsampling replicates samples from the minority class to balance the classes. Some techniques do this by synthesizing new samples that resemble the minority class data while other methods simply add the same minority samples repeatedly.

Hybrid methods
These do a combination of both downsampling and upsampling.

The themis (*https://oreil.ly/MWdh6*) package has recipe steps that can be used to address class imbalance via subsampling. For simple downsampling, we would use:

```
step_downsample(outcome_column_name)
```

Only the training set should be affected by these techniques. The test set or other holdout samples should be left as is when processed using the recipe. For this reason, all of the subsampling steps default the `skip` argument to have a value of TRUE.

Other step functions are row-based as well: `step_filter()`, `step_sample()`, `step_slice()`, and `step_arrange()`. In almost all uses of these steps, the `skip` argument should be set to TRUE.

General Transformations

Mirroring the original dplyr operation, `step_mutate()` can be used to conduct a variety of basic operations on the data. It is best used for straightforward transformations like computing a ratio of two variables, such as `Bedroom_AbvGr / Full_Bath`, the ratio of bedrooms to bathrooms for the Ames housing data.

> When using this flexible step, use extra care to avoid data leakage in your preprocessing. Consider, for example, the transformation `x = w > mean(w)`. When applied to new data or testing data, this transformation would use the mean of w from the *new* data, not the mean of w from the training data.

Natural Language Processing

Recipes can also handle data that are not in the traditional structure where the columns are features. For example, the textrecipes (*https://oreil.ly/iwP9x*) package can apply natural language processing methods to the data. The input column is typically a string of text, and different steps can be used to tokenize the data (e.g., split the text into separate words), filter out tokens, and create new features appropriate for modeling.

Skipping Steps for New Data

The sale price data are already log-transformed in the `ames` data frame. Why not use:

```
step_log(Sale_Price, base = 10)
```

This will cause a failure when the recipe is applied to new properties with an unknown sale price. Since price is what we are trying to predict, there probably won't be a column in the data for this variable. In fact, to avoid information leakage, many tidymodels packages isolate the data being used when making any predictions. This means that the training set and any outcome columns are not available for use at prediction time.

> For simple transformations of the outcome column(s), we strongly suggest that those operations be *conducted outside of the recipe*.

However, there are other circumstances in which this is not an adequate solution. For example, in classification models where there is a severe class imbalance, it is common to conduct *subsampling* of the data that are given to the modeling function. For example, suppose that there were two classes and a 10% event rate. A simple, albeit controversial, approach would be to *downsample* the data so that the model is provided with all of the events and a random 10% of the nonevent samples.

The problem is that the same subsampling process should not be applied to the data being predicted. As a result, when using a recipe, we need a mechanism to ensure that some operations are applied only to the data that are given to the model. Each step function has an option called `skip` that, when set to TRUE, will be ignored by the `predict()` function. In this way, you can isolate the steps that affect the modeling data without causing errors when applied to new samples. However, all steps are applied when using `fit()`.

At the time of writing, the step functions in the recipes and themis packages that are applied only to the training data are:

- `step_adasyn()`
- `step_bsmote()`
- `step_downsample()`
- `step_filter()`
- `step_nearmiss()`
- `step_rose()`
- `step_sample`
- `step_slice()`
- `step_smote()`
- `step_smotenc()`
- `step_tomek()`
- `step_upsample()`

Tidy a recipe()

In Chapter 3, we introduced the `tidy()` verb for statistical objects. There is also a `tidy()` method for recipes, as well as individual recipe steps. Before proceeding, let's create an extended recipe for the Ames data using some of the new steps we've discussed in this chapter:

```
ames_rec <-
  recipe(Sale_Price ~ Neighborhood + Gr_Liv_Area + Year_Built + Bldg_Type +
           Latitude + Longitude, data = ames_train) %>%
  step_log(Gr_Liv_Area, base = 10) %>%
  step_other(Neighborhood, threshold = 0.01) %>%
  step_dummy(all_nominal_predictors()) %>%
  step_interact( ~ Gr_Liv_Area:starts_with("Bldg_Type_") ) %>%
  step_ns(Latitude, Longitude, deg_free = 20)
```

The tidy() method, when called with the recipe object, gives a summary of the recipe steps:

```
tidy(ames_rec)
#> # A tibble: 5 × 6
#>   number operation type     trained skip  id
#>    <int> <chr>     <chr>    <lgl>   <lgl> <chr>
#> 1      1 step      log      FALSE   FALSE log_66JTU
#> 2      2 step      other    FALSE   FALSE other_ePfcw
#> 3      3 step      dummy    FALSE   FALSE dummy_Z18Cl
#> 4      4 step      interact FALSE   FALSE interact_JLU36
#> 5      5 step      ns       FALSE   FALSE ns_rvsqQ
```

This result can be helpful for identifying individual steps, perhaps to then be able to execute the tidy() method on one specific step.

We can specify the id argument in any step function call; otherwise, it is generated using a random suffix. Setting this value can be helpful if the same type of step is added to the recipe more than once. Let's specify the id ahead of time for step_other(), since we'll want to tidy() it:

```
ames_rec <-
  recipe(Sale_Price ~ Neighborhood + Gr_Liv_Area + Year_Built + Bldg_Type +
           Latitude + Longitude, data = ames_train) %>%
  step_log(Gr_Liv_Area, base = 10) %>%
  step_other(Neighborhood, threshold = 0.01, id = "my_id") %>%
  step_dummy(all_nominal_predictors()) %>%
  step_interact( ~ Gr_Liv_Area:starts_with("Bldg_Type_") ) %>%
  step_ns(Latitude, Longitude, deg_free = 20)
```

We'll refit the workflow with this new recipe:

```
lm_wflow <-
  workflow() %>%
  add_model(lm_model) %>%
  add_recipe(ames_rec)

lm_fit <- fit(lm_wflow, ames_train)
```

The tidy() method can be called again along with the id identifier we specified to get our results for applying step_other():

```
estimated_recipe <-
  lm_fit %>%
  extract_recipe(estimated = TRUE)

tidy(estimated_recipe, id = "my_id")
#> # A tibble: 22 × 3
#>   terms        retained             id
#>   <chr>        <chr>                <chr>
#> 1 Neighborhood North_Ames           my_id
#> 2 Neighborhood College_Creek        my_id
#> 3 Neighborhood Old_Town             my_id
#> 4 Neighborhood Edwards              my_id
#> 5 Neighborhood Somerset             my_id
#> 6 Neighborhood Northridge_Heights   my_id
#> # … with 16 more rows
```

The tidy() results we see here for using step_other() show which factor levels were retained—in other words, not added to the new "other" category.

The tidy() method can be called with the number identifier as well, if we know which step in the recipe we need:

```
tidy(estimated_recipe, number = 2)
#> # A tibble: 22 × 3
#>   terms          retained              id
#>   <chr>          <chr>                 <chr>
#> 1 Neighborhood   North_Ames            my_id
#> 2 Neighborhood   College_Creek         my_id
#> 3 Neighborhood   Old_Town              my_id
#> 4 Neighborhood   Edwards               my_id
#> 5 Neighborhood   Somerset              my_id
#> 6 Neighborhood   Northridge_Heights    my_id
#> # … with 16 more rows
```

Each tidy() method returns the relevant information about that step. For example, the tidy() method for step_dummy() returns a column with the variables that were converted to dummy variables and another column with all of the known levels for each column.

Column Roles

When a formula is used with the initial call to recipe(), it assigns *roles* to each of the columns, depending on which side of the tilde they are on. Those roles are either "predictor" or "outcome". However, other roles can be assigned as needed.

For example, in our Ames data set, the original raw data contained a column for address.[1] It may be useful to keep that column in the data so that, after predictions are made, problematic results can be investigated in detail. In other words, the column could be important even when it isn't a predictor or outcome.

To solve this, the add_role(), remove_role(), and update_role() functions can be helpful. For example, for the house price data, the role of the street address column could be modified using:

```
ames_rec %>% update_role(address, new_role = "street address")
```

After this change, the address column in the data frame will no longer be a predictor but instead will be a "street address" according to the recipe. Any character string can be used as a role. Also, columns can have multiple roles (additional roles are added via add_role()) so that they can be selected under more than one context.

This can be helpful when the data are *resampled*. It helps to keep the columns that are not involved with the model fit in the same data frame (rather than in an external

---

1 Our version of these data does not contain that column.

vector). Resampling, described in Chapter 10, creates alternate versions of the data mostly by row subsampling. If the street address were in another column, additional subsampling would be required and might lead to more complex code and a higher likelihood of errors.

Finally, all step functions have a `role` field that can assign roles to the results of the step. In many cases, columns affected by a step retain their existing role. For example, the `step_log()` calls to our `ames_rec` object affected the `Gr_Liv_Area` column. For that step, the default behavior is to keep the existing role for this column since no new column is created. As a counterexample, the step to produce splines defaults new columns to have a role of `"predictor"` since that is usually how spline columns are used in a model. Most steps have sensible defaults, but since the defaults can be different, be sure to check the documentation page to understand which role(s) will be assigned.

Chapter Summary

In this chapter, you learned about using recipes for flexible feature engineering and data preprocessing, from creating dummy variables to handling class imbalance and more. Feature engineering is an important part of the modeling process where information leakage can easily occur and good practices must be adopted. Between the recipes package and other packages that extend recipes, there are over 100 available steps. All possible recipe steps are enumerated at tidymodels website (*https://oreil.ly/FB0BM*). The recipes framework provides a rich data manipulation environment for preprocessing and transforming data prior to modeling. Additionally, you can see how custom steps can be created (*https://oreil.ly/0JPFP*).

Our work here has used recipes solely inside of a workflow object. For modeling, that is the recommended use because feature engineering should be estimated together with a model. However, for visualization and other activities, a workflow may not be appropriate; more recipe-specific functions may be required. Chapter 16 discusses lower-level APIs for fitting, using, and troubleshooting recipes.

The code that we will use in later chapters is:

```
library(tidymodels)
data(ames)
ames <- mutate(ames, Sale_Price = log10(Sale_Price))

set.seed(502)
ames_split <- initial_split(ames, prop = 0.80, strata = Sale_Price)
ames_train <- training(ames_split)
ames_test  <- testing(ames_split)

ames_rec <-
  recipe(Sale_Price ~ Neighborhood + Gr_Liv_Area + Year_Built + Bldg_Type +
           Latitude + Longitude, data = ames_train) %>%
  step_log(Gr_Liv_Area, base = 10) %>%
```

```
  step_other(Neighborhood, threshold = 0.01) %>%
  step_dummy(all_nominal_predictors()) %>%
  step_interact( ~ Gr_Liv_Area:starts_with("Bldg_Type_") ) %>%
  step_ns(Latitude, Longitude, deg_free = 20)

lm_model <- linear_reg() %>% set_engine("lm")

lm_wflow <-
  workflow() %>%
  add_model(lm_model) %>%
  add_recipe(ames_rec)

lm_fit <- fit(lm_wflow, ames_train)
```

## CHAPTER 9

Judging Model Effectiveness

Once we have a model, we need to know how well it works. A quantitative approach for estimating effectiveness allows us to understand the model, to compare different models, or to tweak the model to improve performance. Our focus in tidymodels is on empirical validation; this usually means using data that were not used to create the model as the substrate to measure effectiveness.

The best approach to empirical validation involves using *resampling* methods that will be introduced in Chapter 10. In this chapter, we will motivate the need for empirical validation by using the test set. Keep in mind that the test set can only be used once, as explained in Chapter 5.

When judging model effectiveness, your decision about which metrics to examine can be critical. In later chapters, certain model parameters will be empirically optimized, and a primary performance metric will be used to choose the best submodel. Choosing the wrong metric can easily result in unintended consequences. For example, two common metrics for regression models are the root mean squared error (RMSE) and the coefficient of determination (a.k.a. R^2). The former measures *accuracy* while the latter measures *correlation*. These are not necessarily the same thing. Figure 9-1 demonstrates the difference between the two.

A model optimized for RMSE has more variability but has relatively uniform accuracy across the range of the outcome. The right panel shows that there is a tighter correlation between the observed and predicted values but this model performs poorly in the tails.

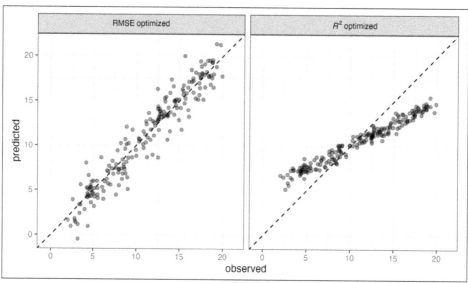

Figure 9-1. Observed versus predicted values for models that are optimized using the RMSE compared to the coefficient of determination.

This chapter will demonstrate the yardstick package, a core tidymodels packages with the focus of measuring model performance. Before illustrating syntax, let's explore whether empirical validation using performance metrics is worthwhile when a model is focused on inference rather than prediction.

Performance Metrics and Inference

The effectiveness of any given model depends on how the model will be used. An inferential model is used primarily to understand relationships, and typically emphasizes the choice (and validity) of probabilistic distributions and other generative qualities that define the model. For a model used primarily for prediction, by contrast, predictive strength is of primary importance and other concerns about underlying statistical qualities may be less important. Predictive strength is usually determined by how close our predictions come to the observed data, i.e., fidelity of the model predictions to the actual results. This chapter focuses on functions that can be used to measure predictive strength. However, our advice for those developing inferential models is to use these techniques even when the model will not be used with the primary goal of prediction.

A longstanding issue with the practice of inferential statistics is that, with a focus purely on inference, it's difficult to assess the credibility of a model. For example, consider the Alzheimer's disease data from Craig-Schapiro et al. (2011) when 333 patients were studied to determine the factors that influence cognitive impairment.

An analysis might take the known risk factors and build a logistic regression model where the outcome is binary (impaired/nonimpaired). Let's consider predictors for age, sex, and the apolipoprotein E genotype. The latter is a categorical variable with six possible combinations of the three main variants of this gene. Apolipoprotein E is known to be associated with dementia (Jungsu, Basak, and Holtzman 2009).

A superficial, but not uncommon, approach to this analysis would be to fit a large model with main effects and interactions, then use statistical tests to find the minimal set of model terms that are statistically significant at some predefined level. If a full model with the three factors and their two- and three-way interactions were used, an initial phase would be to test the interactions using sequential likelihood ratio tests (Hosmer and Lemeshow 2000). Let's step through this kind of approach for the example Alzheimer's disease data:

- When comparing the model with all two-way interactions to one with the additional three-way interaction, the likelihood ratio test produces a p-value of 0.888. This implies that there is no evidence that the four additional model terms associated with the three-way interaction explain enough of the variation in the data to keep them in the model.

- Next, the two-way interactions are similarly evaluated against the model with no interactions. The p-value here is 0.0382. This is somewhat borderline, but, given the small sample size, it would be prudent to conclude that there is evidence that some of the 10 possible two-way interactions are important to the model.

- From here, we would build some explanation of the results. The interactions would be particularly important to discuss since they may spark interesting physiological or neurological hypotheses to be explored further.

While shallow, this analysis strategy is common in practice as well as in the literature. This is especially true if the practitioner has limited formal training in data analysis.

One missing piece of information in this approach is how closely this model fits the actual data. Using resampling methods, discussed in Chapter 10, we can estimate the accuracy of this model to be about 73.3%. Accuracy is often a poor measure of model performance; we use it here because it is commonly understood. If the model has 73.3% fidelity to the data, should we trust conclusions it produces? We might think so until we realize that the baseline rate of nonimpaired patients in the data is 72.7%. This means that, despite our statistical analysis, the two-factor model appears to be only 0.6% better than a simple heuristic that always predicts patients to be unimpaired, regardless of the observed data.

In the remainder of this chapter, we will discuss general approaches for evaluating models via empirical validation. These approaches are grouped by the nature of the outcome data: purely numeric, binary classes, and three or more class levels.

The point of this analysis is to demonstrate the idea that optimization of statistical characteristics of the model does not imply that the model fits the data well. Even for purely inferential models, some measure of fidelity to the data should accompany the inferential results. Using this, the consumers of the analyses can calibrate their expectations of the results.

Regression Metrics

Recall from Chapter 6 that tidymodels prediction functions produce tibbles with columns for the predicted values. These columns have consistent names, and the functions in the yardstick package that produce performance metrics have consistent interfaces. The functions are data frame-based, as opposed to vector-based, with the general syntax of:

```
function(data, truth, ...)
```

where `data` is a data frame or tibble and `truth` is the column with the observed outcome values. The ellipses or other arguments are used to specify the column(s) containing the predictions.

To illustrate, let's take the model from the very end of Chapter 8. This model `lm_wflow_fit` combines a linear regression model with a predictor set supplemented with an interaction and spline functions for longitude and latitude. It was created from a training set (named `ames_train`). Although we do not advise using the test set at this juncture of the modeling process, it will be used here to illustrate functionality and syntax. The data frame `ames_test` consists of 588 properties. To start, let's produce predictions:

```
ames_test_res <- predict(lm_fit, new_data = ames_test %>% select(-Sale_Price))
ames_test_res
#> # A tibble: 588 × 1
#>    .pred
#>    <dbl>
#> 1  5.07
#> 2  5.31
#> 3  5.28
#> 4  5.33
#> 5  5.30
#> 6  5.24
#> # … with 582 more rows
```

The predicted numeric outcome from the regression model is named `.pred`. Let's match the predicted values with their corresponding observed outcome values:

```
ames_test_res <- bind_cols(ames_test_res, ames_test %>% select(Sale_Price))
ames_test_res
#> # A tibble: 588 × 2
#>    .pred Sale_Price
#>    <dbl>      <dbl>
#> 1  5.07       5.02
```

```
#> 2  5.31      5.39
#> 3  5.28      5.28
#> 4  5.33      5.28
#> 5  5.30      5.28
#> 6  5.24      5.26
#> # … with 582 more rows
```

We see that these values mostly look close, but we don't yet have a quantitative under-standing of how the model is doing because we haven't computed any performance metrics. Note that both the predicted and observed outcomes are in log-10 units. It is best practice to analyze the predictions on the transformed scale (if one were used), even if the predictions are reported using the original units.

Let's plot the data in Figure 9-2 before computing metrics:

```
ggplot(ames_test_res, aes(x = Sale_Price, y = .pred)) +
  # Create a diagonal line:
  geom_abline(lty = 2) +
  geom_point(alpha = 0.5) +
  labs(y = "Predicted Sale Price (log10)", x = "Sale Price (log10)") +
  # Scale and size the x- and y-axis uniformly:
  coord_obs_pred()
```

There is one low-price property that is substantially overpredicted, i.e., quite high above the dashed line.

Let's compute the root mean squared error for this model using the rmse() function:

```
rmse(ames_test_res, truth = Sale_Price, estimate = .pred)
#> # A tibble: 1 × 3
#>   .metric .estimator .estimate
#>   <chr>   <chr>          <dbl>
#> 1 rmse    standard      0.0736
```

This shows us the standard format of the output of yardstick functions. Metrics for numeric outcomes usually have a value of "standard" for the .estimator column. Examples with different values for this column are shown in the next sections.

To compute multiple metrics at once, we can create a *metric set*. Let's add R^2 and the mean absolute error:

```
ames_metrics <- metric_set(rmse, rsq, mae)
ames_metrics(ames_test_res, truth = Sale_Price, estimate = .pred)
#> # A tibble: 3 × 3
#>   .metric .estimator .estimate
#>   <chr>   <chr>          <dbl>
#> 1 rmse    standard      0.0736
#> 2 rsq     standard      0.836
#> 3 mae     standard      0.0549
```

This tidy data format stacks the metrics vertically. The root mean squared error and mean absolute error metrics are both on the scale of the outcome (so log10(Sale_Price) for our example) and measure the difference between the predicted and observed values. The value for R^2 measures the squared correlation between the predicted and observed values, so values closer to one are better.

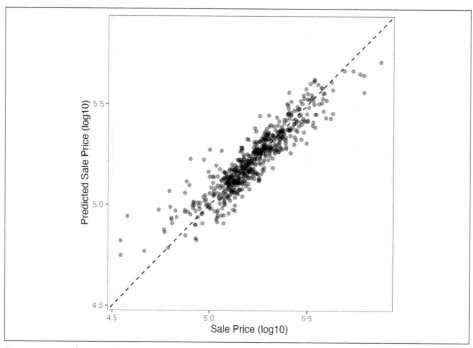

Figure 9-2. Observed versus predicted values for an Ames regression model, with log-10 units on both axes.

 The yardstick package does *not* contain a function for adjusted R^2. This modification of the coefficient of determination is commonly used when the same data used to fit the model are used to evaluate the model. This metric is not fully supported in tidymodels because it is always a better approach to compute performance on a separate data set than the one used to fit the model.

Binary Classification Metrics

To illustrate other ways to measure model performance, we will switch to a different example. The modeldata package (another one of the tidymodels packages) contains example predictions from a test data set with two classes ("Class1" and "Class2"):

```
data(two_class_example)
tibble(two_class_example)
#> # A tibble: 500 × 4
#>    truth   Class1   Class2 predicted
#>    <fct>    <dbl>    <dbl> <fct>
#> 1 Class2 0.00359 0.996    Class2
#> 2 Class1 0.679   0.321    Class1
#> 3 Class2 0.111   0.889    Class2
#> 4 Class1 0.735   0.265    Class1
#> 5 Class2 0.0162  0.984    Class2
```

```
#> 6 Class1 0.999    0.000725 Class1
#> # … with 494 more rows
```

The second and third columns are the predicted class probabilities for the test set while predicted are the discrete predictions.

For the hard class predictions, a variety of yardstick functions are helpful:

```
# A confusion matrix:
conf_mat(two_class_example, truth = truth, estimate = predicted)
#>           Truth
#> Prediction Class1 Class2
#>     Class1    227     50
#>     Class2     31    192

# Accuracy:
accuracy(two_class_example, truth, predicted)
#> # A tibble: 1 × 3
#>   .metric  .estimator .estimate
#>   <chr>    <chr>          <dbl>
#> 1 accuracy binary         0.838

# Matthews correlation coefficient:
mcc(two_class_example, truth, predicted)
#> # A tibble: 1 × 3
#>   .metric .estimator .estimate
#>   <chr>   <chr>          <dbl>
#> 1 mcc     binary         0.677

# F1 metric:
f_meas(two_class_example, truth, predicted)
#> # A tibble: 1 × 3
#>   .metric .estimator .estimate
#>   <chr>   <chr>          <dbl>
#> 1 f_meas  binary         0.849

# Combining these three classification metrics together
classification_metrics <- metric_set(accuracy, mcc, f_meas)
classification_metrics(two_class_example, truth = truth, estimate = predicted)
#> # A tibble: 3 × 3
#>   .metric  .estimator .estimate
#>   <chr>    <chr>          <dbl>
#> 1 accuracy binary         0.838
#> 2 mcc      binary         0.677
#> 3 f_meas   binary         0.849
```

The Matthews correlation coefficient and F1 score both summarize the confusion matrix, but compared to mcc(), which measures the quality of both positive and negative examples, the f_meas() metric emphasizes the positive class, i.e., the event of interest. For binary classification data sets like this example, yardstick functions have a standard argument called event_level to distinguish positive and negative levels. The default (which we used in this code) is that the *first* level of the outcome factor is the event of interest.

There is some heterogeneity in R functions in this regard; some use the first level and others the second to denote the event of interest. We consider it more intuitive that the first level is the most important. The second-level logic is borne of encoding the outcome as 0/1 (in which case the second value is the event) and unfortunately remains in some packages. However, tidymodels (along with many other R packages) requires a categorical outcome to be encoded as a factor and, for this reason, the legacy justification for the second level as the event becomes irrelevant.

As an example where the second level is the event:

```
f_meas(two_class_example, truth, predicted, event_level = "second")
#> # A tibble: 1 × 3
#>   .metric .estimator .estimate
#>   <chr>   <chr>          <dbl>
#> 1 f_meas  binary         0.826
```

In this output, the .estimator value of "binary" indicates that the standard formula for binary classes will be used.

There are numerous classification metrics that use the predicted probabilities as inputs rather than the hard class predictions. For example, the receiver operating characteristic (ROC) curve computes the sensitivity and specificity over a continuum of different event thresholds. The predicted class column is not used. There are two yardstick functions for this method: roc_curve() computes the data points that make up the ROC curve and roc_auc() computes the area under the curve.

The interfaces to these types of metric functions use the ... argument placeholder to pass in the appropriate class probability column. For two-class problems, the probability column for the event of interest is passed into the function:

```
two_class_curve <- roc_curve(two_class_example, truth, Class1)
two_class_curve
#> # A tibble: 502 × 3
#>   .threshold specificity sensitivity
#>        <dbl>       <dbl>       <dbl>
#> 1 -Inf         0                 1
#> 2  1.79e-7     0                 1
#> 3  4.50e-6     0.00413           1
#> 4  5.81e-6     0.00826           1
#> 5  5.92e-6     0.0124            1
#> 6  1.22e-5     0.0165            1
#> # … with 496 more rows

roc_auc(two_class_example, truth, Class1)
#> # A tibble: 1 × 3
#>   .metric .estimator .estimate
#>   <chr>   <chr>          <dbl>
#> 1 roc_auc binary         0.939
```

The `two_class_curve` object can be used in a `ggplot` call to visualize the curve, as shown in Figure 9-3. There is an `autoplot()` method that will take care of the details:

```
autoplot(two_class_curve)
```

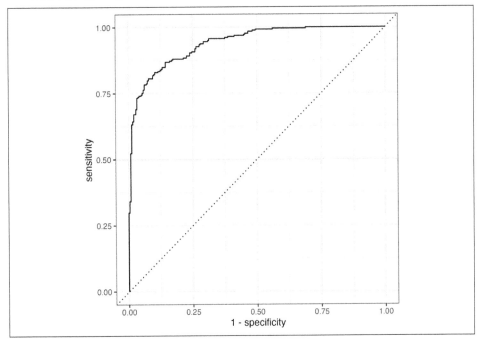

Figure 9-3. Example ROC curve.

If the curve was close to the diagonal line, then the model's predictions would be no better than random guessing. Since the curve is up in the top left-hand corner, we see that our model performs well at different thresholds.

There are a number of other functions that use probability estimates, including `gain_curve()`, `lift_curve()`, and `pr_curve()`.

Multiclass Classification Metrics

What about data with three or more classes? To demonstrate, let's explore a different example data set that has four classes:

```
data(hpc_cv)
tibble(hpc_cv)
#> # A tibble: 3,467 × 7
#>    obs   pred     VF      F       M            L Resample
#>    <fct> <fct> <dbl>  <dbl>   <dbl>        <dbl> <chr>
#> 1 VF    VF    0.914 0.0779 0.00848 0.0000199    Fold01
#> 2 VF    VF    0.938 0.0571 0.00482 0.0000101    Fold01
#> 3 VF    VF    0.947 0.0495 0.00316 0.00000500   Fold01
```

```
#> 4 VF     VF     0.929 0.0653 0.00579 0.0000156  Fold01
#> 5 VF     VF     0.942 0.0543 0.00381 0.00000729 Fold01
#> 6 VF     VF     0.951 0.0462 0.00272 0.00000384 Fold01
#> # … with 3,461 more rows
```

As before, there are factors for the observed and predicted outcomes along with four other columns of predicted probabilities for each class. (These data also include a Resample column. These hpc_cv results are for out-of-sample predictions associated with 10-fold cross-validation. For the time being, this column will be ignored and we'll discuss resampling in depth in Chapter 10.)

The functions for metrics that use the discrete class predictions are identical to their binary counterparts:

```
accuracy(hpc_cv, obs, pred)
#> # A tibble: 1 × 3
#>   .metric  .estimator .estimate
#>   <chr>    <chr>          <dbl>
#> 1 accuracy multiclass     0.709

mcc(hpc_cv, obs, pred)
#> # A tibble: 1 × 3
#>   .metric .estimator .estimate
#>   <chr>   <chr>          <dbl>
#> 1 mcc     multiclass     0.515
```

Note that, in these results, a "multiclass" .estimator is listed. Like "binary," this indicates that the formula for outcomes with three or more class levels was used. The Matthews correlation coefficient was originally designed for two classes but has been extended to cases with more class levels.

There are methods for taking metrics designed to handle outcomes with only two classes and extend them for outcomes with more than two classes. For example, a metric such as sensitivity measures the true positive rate which, by definition, is specific to two classes (i.e., "event" and "nonevent"). How can this metric be used in our example data?

There are wrapper methods that can be used to apply sensitivity to our four-class outcome. These options are macro-averaging, macro-weighted averaging, and micro-averaging:

- Macro-averaging computes a set of one-versus-all metrics using the standard two-class statistics. These are averaged.

- Macro-weighted averaging does the same, but the average is weighted by the number of samples in each class.

- Micro-averaging computes the contribution for each class, aggregates them, then computes a single metric from the aggregates.

See Wu and Zhou (2017) and Opitz and Burst (2019) for more on extending classification metrics to outcomes with more than two classes.

Using sensitivity as an example, the usual two-class calculation is the ratio of the
number of correctly predicted events divided by the number of true events. The
manual calculations for these averaging methods are:

```
class_totals <-
  count(hpc_cv, obs, name = "totals") %>%
  mutate(class_wts = totals / sum(totals))
class_totals
#>   obs totals class_wts
#> 1  VF   1769   0.51024
#> 2   F   1078   0.31093
#> 3   M    412   0.11883
#> 4   L    208   0.05999

cell_counts <-
  hpc_cv %>%
  group_by(obs, pred) %>%
  count() %>%
  ungroup()

# Compute the four sensitivities using 1-vs-all
one_versus_all <-
  cell_counts %>%
  filter(obs == pred) %>%
  full_join(class_totals, by = "obs") %>%
  mutate(sens = n / totals)
one_versus_all
#> # A tibble: 4 × 6
#>   obs   pred      n totals class_wts  sens
#>   <fct> <fct> <int>  <int>     <dbl> <dbl>
#> 1 VF    VF     1620   1769    0.510  0.916
#> 2 F     F       647   1078    0.311  0.600
#> 3 M     M        79    412    0.119  0.192
#> 4 L     L       111    208    0.0600 0.534

# Three different estimates:
one_versus_all %>%
  summarize(
    macro = mean(sens),
    macro_wts = weighted.mean(sens, class_wts),
    micro = sum(n) / sum(totals)
  )
#> # A tibble: 1 × 3
#>   macro macro_wts micro
#>   <dbl>     <dbl> <dbl>
#> 1 0.560     0.709 0.709
```

Thankfully, there is no need to manually implement these averaging methods.
Instead, yardstick functions can automatically apply these methods via the estimator
argument:

```
sensitivity(hpc_cv, obs, pred, estimator = "macro")
#> # A tibble: 1 × 3
#>   .metric     .estimator .estimate
```

```
#>    <chr>        <chr>            <dbl>
#> 1 sensitivity macro            0.560
sensitivity(hpc_cv, obs, pred, estimator = "macro_weighted")
#> # A tibble: 1 × 3
#>    .metric    .estimator      .estimate
#>    <chr>       <chr>             <dbl>
#> 1 sensitivity macro_weighted   0.709
sensitivity(hpc_cv, obs, pred, estimator = "micro")
#> # A tibble: 1 × 3
#>    .metric    .estimator .estimate
#>    <chr>       <chr>         <dbl>
#> 1 sensitivity micro         0.709
```

When dealing with probability estimates, there are some metrics with multiclass analogs. For example, Hand and Till (2001) determined a multiclass technique for ROC curves. In this case, *all* of the class probability columns must be given to the function:

```
roc_auc(hpc_cv, obs, VF, F, M, L)
#> # A tibble: 1 × 3
#>    .metric .estimator .estimate
#>    <chr>   <chr>         <dbl>
#> 1 roc_auc hand_till     0.829
```

Macro-weighted averaging is also available as an option for applying this metric to a multiclass outcome:

```
roc_auc(hpc_cv, obs, VF, F, M, L, estimator = "macro_weighted")
#> # A tibble: 1 × 3
#>    .metric .estimator      .estimate
#>    <chr>   <chr>             <dbl>
#> 1 roc_auc macro_weighted    0.868
```

Finally, all of these performance metrics can be computed using dplyr groupings. Recall that these data have a column for the resampling groups. We haven't yet discussed resampling in detail, but notice how we can pass a grouped data frame to the metric function to compute the metrics for each group:

```
hpc_cv %>%
  group_by(Resample) %>%
  accuracy(obs, pred)
#> # A tibble: 10 × 4
#>    Resample .metric  .estimator .estimate
#>    <chr>    <chr>    <chr>         <dbl>
#> 1 Fold01   accuracy multiclass   0.726
#> 2 Fold02   accuracy multiclass   0.712
#> 3 Fold03   accuracy multiclass   0.758
#> 4 Fold04   accuracy multiclass   0.712
#> 5 Fold05   accuracy multiclass   0.712
#> 6 Fold06   accuracy multiclass   0.697
#> # … with 4 more rows
```

The groupings also translate to the `autoplot()` methods, with results shown in Figure 9-4:

```
# Four 1-vs-all ROC curves for each fold
hpc_cv %>%
  group_by(Resample) %>%
  roc_curve(obs, VF, F, M, L) %>%
  autoplot() +
  theme(legend.position = "none")
```

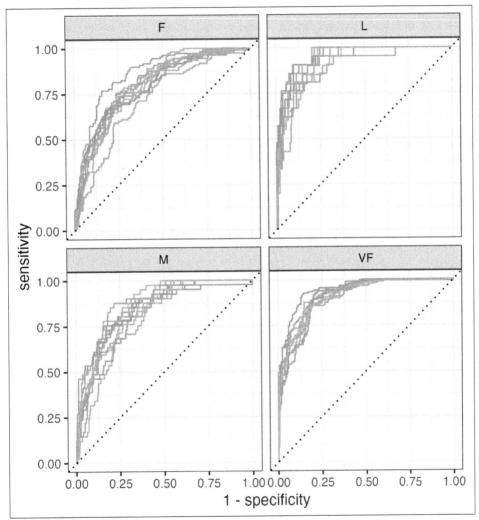

Figure 9-4. Resampled ROC curves for each of the four outcome classes.

This visualization shows us that the different groups all perform about the same, but that the VF class is predicted better than the F or M classes, since the VF ROC curves are more in the top-left corner. This example uses resamples as the groups, but any grouping in your data can be used. This `autoplot()` method can be a quick visualization method for model effectiveness across outcome classes and/or groups.

Chapter Summary

Different metrics measure different aspects of a model fit; foor example, RMSE measures accuracy while the R^2 measures correlation. Measuring model performance is important even when a given model will not be used primarily for prediction; predictive power is also important for inferential or descriptive models. Functions from the yardstick package measure the effectiveness of a model using data. The primary tidymodels interface uses tidyverse principles and data frames (as opposed to having vector arguments). Different metrics are appropriate for regression and classification metrics and, within these, there are sometimes different ways to estimate the statistics, such as for multiclass outcomes.

## PART III

Tools for Creating Effective Models

## CHAPTER 10

Resampling for Evaluating Performance

We have already covered several pieces that must be put together to evaluate the performance of a model. Chapter 9 described statistics for measuring model performance. Chapter 5 introduced the idea of data spending, and we recommended the test set for obtaining an unbiased estimate of performance. However, we usually need to understand the performance of a model or even multiple models *before using the test set*.

Typically we can't decide on which final model to use with the test set before first assessing model performance. There is a gap between our need to measure performance reliably and the data splits (training and testing) we have available.

In this chapter, we describe an approach called resampling that can fill this gap. Resampling estimates of performance can generalize to new data in a similar way as estimates from a test set. Chapter 11 complements this one by demonstrating statistical methods that compare resampling results.

In order to fully appreciate the value of resampling, let's first take a look the resubstitution approach, which can often fail.

The Resubstitution Approach

When we measure performance on the same data that we used for training (as opposed to new data or testing data), we say we have *resubstituted* the data. Let's again use the Ames data to demonstrate these concepts. The end of Chapter 8 summarizes the current state of our Ames analysis. It includes a recipe object named `ames_rec`, a linear model, and a workflow using that recipe and model called lm_wflow. This workflow was fit on the training set, resulting in lm_fit.

For a comparison to this linear model, we can also fit a different type of model. *Random forests* are a tree ensemble method that operates by creating a large number of decision trees from slightly different versions of the training set (Breiman 2001a). This collection of trees makes up the ensemble. When predicting a new sample, each ensemble member makes a separate prediction. These are averaged to create the final ensemble prediction for the new data point.

Random forest models are very powerful, and they can emulate the underlying data patterns very closely. While this model can be computationally intensive, it is very low maintenance; very little preprocessing is required (as documented in the Appendix).

Using the same predictor set as the linear model (without the extra preprocessing steps), we can fit a random forest model to the training set via the "ranger" engine (which uses the ranger R package for computation). This model requires no preprocessing, so a simple formula can be used:

```
rf_model <-
  rand_forest(trees = 1000) %>%
  set_engine("ranger") %>%
  set_mode("regression")

rf_wflow <-
  workflow() %>%
  add_formula(
    Sale_Price ~ Neighborhood + Gr_Liv_Area + Year_Built + Bldg_Type +
      Latitude + Longitude) %>%
  add_model(rf_model)

rf_fit <- rf_wflow %>% fit(data = ames_train)
```

How should we compare the linear and random forest models? For demonstration, we will predict the training set to produce what is known as an *apparent metric* or *resubstitution metric*. This function creates predictions and formats the results:

```
estimate_perf <- function(model, dat) {
  # Capture the names of the `model` and `dat` objects
  cl <- match.call()
  obj_name <- as.character(cl$model)
  data_name <- as.character(cl$dat)
  data_name <- gsub("ames_", "", data_name)

  # Estimate these metrics:
  reg_metrics <- metric_set(rmse, rsq)

  model %>%
    predict(dat) %>%
    bind_cols(dat %>% select(Sale_Price)) %>%
    reg_metrics(Sale_Price, .pred) %>%
    select(-.estimator) %>%
```

```
    mutate(object = obj_name, data = data_name)
}
```

Both RMSE and R^2 are computed. The resubstitution statistics are:

```
estimate_perf(rf_fit, ames_train)
#> # A tibble: 2 × 4
#>   .metric .estimate object data
#>   <chr>       <dbl> <chr>  <chr>
#> 1 rmse       0.0365 rf_fit train
#> 2 rsq        0.960  rf_fit train
estimate_perf(lm_fit, ames_train)
#> # A tibble: 2 × 4
#>   .metric .estimate object data
#>   <chr>       <dbl> <chr>  <chr>
#> 1 rmse       0.0754 lm_fit train
#> 2 rsq        0.816  lm_fit train
```

Based on these results, the random forest is much more capable of predicting the sale prices; the RMSE estimate is two-fold better than linear regression. If we needed to choose between these two models for this price prediction problem, we would probably chose the random forest because, on the log scale we are using, its RMSE is about half as large. The next step applies the random forest model to the test set for final verification:

```
estimate_perf(rf_fit, ames_test)
#> # A tibble: 2 × 4
#>   .metric .estimate object data
#>   <chr>       <dbl> <chr>  <chr>
#> 1 rmse       0.0704 rf_fit test
#> 2 rsq        0.852  rf_fit test
```

The test set RMSE estimate, 0.0704, is *much worse than the training set* value of 0.0365! Why did this happen?

Many predictive models are capable of learning complex trends from the data. In statistics, these are commonly referred to as *low bias models*.

In this context, *bias* is the difference between the true pattern or relationships in data and the types of patterns that the model can emulate. Many black-box machine learning models have low bias, meaning they can reproduce complex relationships. Other models (such as linear/logistic regression, discriminant analysis, and others) are not as adaptable and are considered *high bias* models.[1]

For a low bias model, the high degree of predictive capacity can sometimes result in the model nearly memorizing the training set data. As an obvious example, consider a KNN model with only one neighbor. It will always provide perfect predictions for the training set no matter how well it truly works for other data sets. Random forest

---

1 See section 1.2.5 of Kuhn and Johnson (2020) (*https://oreil.ly/pfcLQ*) for a discussion.

models are similar; repredicting the training set will always result in an artificially optimistic estimate of performance.

For both models, Table 10-1 summarizes the RMSE estimate for the training and test sets:

Table 10-1. Performance statistics for training and test sets.

object	Train	Test
lm_fit	0.0754	0.0736
rf_fit	0.0365	0.0704

Notice that the linear regression model is consistent between training and testing, because of its limited complexity.[2]

 The main takeaway from this example is that repredicting the training set will result in an artificially optimistic estimate of performance. It is a bad idea for most models.

If the test set should not be used immediately, and repredicting the training set is a bad idea, what should be done? Resampling methods, such as cross-validation or validation sets, are the solution.

Resampling Methods

Resampling methods are empirical simulation systems that emulate the process of using some data for modeling and different data for evaluation. Most resampling methods are iterative, meaning that this process is repeated multiple times. The diagram in Figure 10-1 illustrates how resampling methods generally operate.

Resampling is conducted only on the training set, as you see in Figure 10-1. The test set is not involved. For each iteration of resampling, the data are partitioned into two subsamples:

- The model is fit with the *analysis set*.
- The model is evaluated with the *assessment set*.

---

2 It is possible for a linear model to nearly memorize the training set, like the random forest model did. In the ames_rec object, change the number of spline terms for longitude and latitude to a large number (say, 1,000). This would produce a model fit with a very small resubstitution RMSE and a test set RMSE that is much larger.

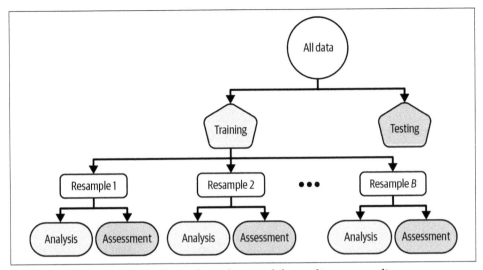

Figure 10-1. Data splitting scheme from the initial data split to resampling.

These two subsamples are somewhat analogous to training and test sets. Our language of *analysis* and *assessment* avoids confusion with the initial split of the data. These data sets are mutually exclusive. The partitioning scheme used to create the analysis and assessment sets is usually the defining characteristic of the method.

Suppose 20 iterations of resampling are conducted. This means that 20 separate models are fit on the analysis sets, and the corresponding assessment sets produce 20 sets of performance statistics. The final estimate of performance for a model is the average of the 20 replicates of the statistics. This average has very good generalization properties and is far better than the resubstitution estimates.

The next section defines several commonly used resampling methods and discusses their pros and cons.

Cross-Validation

Cross-validation is a well established resampling method. While there are a number of variations, the most common cross-validation method is V-fold cross-validation. The data are randomly partitioned into V sets of roughly equal size (called the *folds*). For illustration, $V = 3$ is shown in Figure 10-2 for a data set of 30 training set points with random fold allocations. The number inside the symbols is the sample number.

The shade of the symbols in Figure 10-2 represents their randomly assigned folds. Stratified sampling is also an option for assigning folds (previously discussed in Chapter 5).

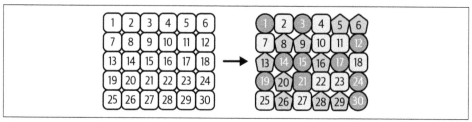

Figure 10-2. V-fold cross-validation randomly assigns data to folds.

For three-fold cross-validation, the three iterations of resampling are illustrated in Figure 10-3. For each iteration, one fold is held out for assessment statistics and the remaining folds are substrate for the model. This process continues for each fold so that three models produce three sets of performance statistics.

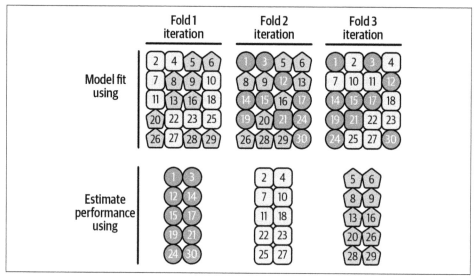

Figure 10-3. V-fold cross-validation data usage.

When $V = 3$, the analysis sets are 2/3 of the training set and each assessment set is a distinct 1/3. The final resampling estimate of performance averages each of the V replicates.

Using $V = 3$ is a good choice to illustrate cross-validation, but it is a poor choice in practice because it is too low to generate reliable estimates. In practice, values of V are most often 5 or 10; we generally prefer 10-fold cross-validation as a default because it is large enough for good results in most situations.

What are the effects of changing V? Larger values result in resampling estimates with small bias but substantial variance. Smaller values of V have large bias but low variance. We prefer 10-fold since noise is reduced by replication, but bias is not.[3]

The primary input is the training set data frame as well as the number of folds (defaulting to 10):

```
set.seed(1001)
ames_folds <- vfold_cv(ames_train, v = 10)
ames_folds
#> #  10-fold cross-validation
#> # A tibble: 10 × 2
#>    splits            id
#>    <list>            <chr>
#> 1 <split [2107/235]> Fold01
#> 2 <split [2107/235]> Fold02
#> 3 <split [2108/234]> Fold03
#> 4 <split [2108/234]> Fold04
#> 5 <split [2108/234]> Fold05
#> 6 <split [2108/234]> Fold06
#> # … with 4 more rows
```

The column named `splits` contains the information on how to split the data (similar to the object used to create the initial training/test partition). While each row of `splits` has an embedded copy of the entire training set, R is smart enough not to make copies of the data in memory.[4] The print method inside of the tibble shows the frequency of each: `[2107/235]` indicates that roughly two thousand samples are in the analysis set and 235 are in that particular assessment set.

These objects also always contain a character column called `id` that labels the partition.[5]

To manually retrieve the partitioned data, the `analysis()` and `assessment()` functions return the corresponding data frames:

```
# For the first fold:
ames_folds$splits[[1]] %>% analysis() %>% dim()
#> [1] 2107   74
```

The tidymodels packages, such as tune (*https://oreil.ly/WdI3T*), contain high-level user interfaces so that functions like `analysis()` are not generally needed for day-to-day work. Chapter 10 demonstrates functions to fit a model over these resamples.

---

3 See section 3.4 of Kuhn and Johnson (2020) (*https://oreil.ly/Mvv6Y*) for a longer description of the results of changing V.

4 To see this for yourself, try executing `lobstr::obj_size(ames_folds)` and `lobstr::obj_size(ames_train)`. The size of the resample object is much less than 10 times the size of the original data.

5 Some resampling methods require multiple `id` fields.

There are a variety of cross-validation variations; we'll go through the most important ones.

Repeated Cross-Validation

The most important variation on cross-validation is repeated V-fold cross-validation. Depending on data size or other characteristics, the resampling estimate produced by V-fold cross-validation may be excessively noisy.[6] As with many statistical problems, one way to reduce noise is to gather more data. For cross-validation, this means averaging more than V statistics.

To create R repeats of V-fold cross-validation, the same fold generation process is done R times to generate R collections of V partitions. Now, instead of averaging V statistics, $V \times R$ statistics produce the final resampling estimate. Due to the Central Limit Theorem, the summary statistics from each model tend toward a normal distribution, as long as we have a lot of data relative to $V \times R$.

Consider the Ames data. On average, 10-fold cross-validation uses assessment sets that contain roughly 234 properties. If RMSE is the statistic of choice, we can denote that estimate's standard deviation as σ. With simple 10-fold cross-validation, the standard error of the mean RMSE is $\sigma/\sqrt{10}$. If this is too noisy, repeats reduce the standard error to $\sigma/\sqrt{10R}$. For 10-fold cross-validation with R replicates, the plot in Figure 10-4 shows how quickly the standard error[7] decreases with replicates.

Larger numbers of replicates tend to have less impact on the standard error. However, if the baseline value of σ is impractically large, the diminishing returns on replication may still be worth the extra computational costs.

To create repeats, invoke `vfold_cv()` with an additional argument `repeats`:

```
vfold_cv(ames_train, v = 10, repeats = 5)
#> #  10-fold cross-validation repeated 5 times
#> # A tibble: 50 × 3
#>    splits             id      id2
#>    <list>             <chr>   <chr>
#> 1 <split [2107/235]> Repeat1 Fold01
#> 2 <split [2107/235]> Repeat1 Fold02
#> 3 <split [2108/234]> Repeat1 Fold03
#> 4 <split [2108/234]> Repeat1 Fold04
#> 5 <split [2108/234]> Repeat1 Fold05
#> 6 <split [2108/234]> Repeat1 Fold06
#> # … with 44 more rows
```

---

6 For more details, see section 3.4.6 of Kuhn and Johnson (2020) (*https://oreil.ly/nt1SS*).

7 These are *approximate* standard errors. As will be discussed in the next chapter, there is a within-replicate correlation that is typical of resampled results. By ignoring this extra component of variation, the simple calculations shown in this plot are overestimates of the reduction in noise in the standard errors.

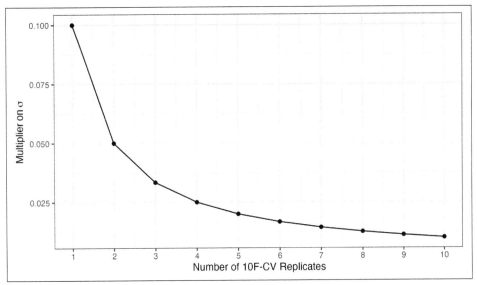

Figure 10-4. Relationship between the relative variance in performance estimates versus the number of cross-validation repeats.

Leave-One-Out Cross-Validation

One variation of cross-validation is leave-one-out (LOO) cross-validation, where V is the number of data points in the training set. If there are n training set samples, n models are fit using $n - 1$ rows of the training set. Each model predicts the single excluded data point. At the end of resampling, the n predictions are pooled to produce a single performance statistic.

Leave-one-out methods are deficient compared to almost any other method. For anything but pathologically small samples, LOO is computationally excessive, and it may not have good statistical properties. Although the rsample package contains a `loo_cv()` function, these objects are not generally integrated into the broader tidymodels frameworks.

Monte Carlo Cross-Validation

Another variant of V-fold cross-validation is Monte Carlo cross-validation (MCCV, Xu and Liang, 2001). Like V-fold cross-validation, it allocates a fixed proportion of data to the assessment sets. The difference between MCCV and regular cross-validation is that, for MCCV, this proportion of the data is randomly selected each time. This results in assessment sets that are not mutually exclusive. To create these resampling objects:

```
mc_cv(ames_train, prop = 9/10, times = 20)
#> # Monte Carlo cross-validation (0.9/0.1) with 20 resamples
#> # A tibble: 20 × 2
#>    splits            id
#>    <list>            <chr>
#> 1 <split [2107/235]> Resample01
#> 2 <split [2107/235]> Resample02
#> 3 <split [2107/235]> Resample03
#> 4 <split [2107/235]> Resample04
#> 5 <split [2107/235]> Resample05
#> 6 <split [2107/235]> Resample06
#> # … with 14 more rows
```

Validation Sets

In Chapter 5, we briefly discussed the use of a validation set, a single partition that is set aside to estimate performance separate from the test set. When using a validation set, the initial available data set is split into a training set, a validation set, and a test set (see Figure 10-5).

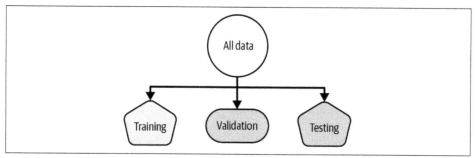

Figure 10-5. A three-way initial split into training, testing, and validation sets.

Validation sets are often used when the original pool of data is very large. In this case, a single large partition may be adequate to characterize model performance without having to do multiple resampling iterations.

With the rsample package, a validation set is like any other resampling object; this type is different only in that it has a single iteration.[8] Figure 10-6 shows this scheme.

To create a validation set object that uses 3/4 of the data for model fitting:

```
set.seed(1002)
val_set <- validation_split(ames_train, prop = 3/4)
val_set
#> # Validation Set Split (0.75/0.25)
#> # A tibble: 1 × 2
#>    splits            id
#>    <list>            <chr>
#> 1 <split [1756/586]> validation
```

---

8 In essence, a validation set can be considered a single iteration of Monte Carlo cross-validation.

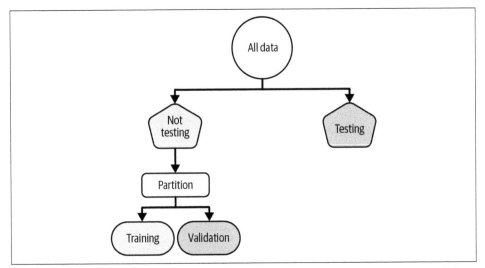

Figure 10-6. A two-way initial split into training and testing with an additional validation set split on the training set.

Bootstrapping

Bootstrap resampling was originally invented as a method for approximating the sampling distribution of statistics whose theoretical properties are intractable (Davison and Hinkley 1997). Using it to estimate model performance is a secondary application of the method.

A bootstrap sample of the training set is a sample that is the same size as the training set but is drawn *with replacement*. This means that some training set data points are selected multiple times for the analysis set. Each data point has a 63.2% chance of inclusion in the training set at least once. The assessment set contains all of the training set samples that were not selected for the analysis set (on average, with 36.8% of the training set). When bootstrapping, the assessment set is often called the *out-of-bag* sample.

For a training set of 30 samples, a schematic of three bootstrap samples is shown in Figure 10-7. Note that the sizes of the assessment sets vary. Using the rsample package, we can create such bootstrap resamples:

```
bootstraps(ames_train, times = 5)
#> # Bootstrap sampling
#> # A tibble: 5 × 2
#>   splits            id
#>   <list>            <chr>
#> 1 <split [2342/858]> Bootstrap1
#> 2 <split [2342/855]> Bootstrap2
#> 3 <split [2342/852]> Bootstrap3
#> 4 <split [2342/851]> Bootstrap4
#> 5 <split [2342/867]> Bootstrap5
```

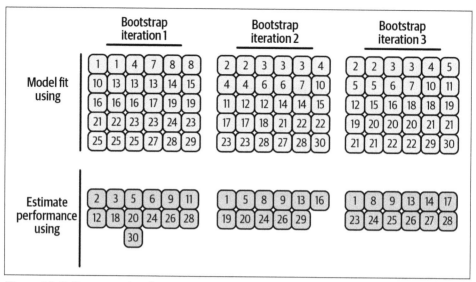

Figure 10-7. Bootstrapping data usage.

Bootstrap samples produce performance estimates that have very low variance (unlike cross-validation) but have significant pessimistic bias. This means that, if the true accuracy of a model is 90%, the bootstrap would tend to estimate the value to be less than 90%. The amount of bias cannot be empirically determined with sufficient accuracy. Additionally, the amount of bias changes over the scale of the performance metric. For example, the bias is likely to be different when the accuracy is 90% versus when it is 70%.

The bootstrap is also used inside of many models. For example, the random forest model mentioned earlier contained 1,000 individual decision trees. Each tree was the product of a different bootstrap sample of the training set.

Rolling Forecasting Origin Resampling

When the data have a strong time component, a resampling method should support modeling to estimate seasonal and other temporal trends within the data. A technique that randomly samples values from the training set can disrupt the model's ability to estimate these patterns.

Rolling forecast origin resampling (Hyndman and Athanasopoulos 2018) provides a method that emulates how time series data are often partitioned in practice, estimating the model with historical data and evaluating it with the most recent data. For this type of resampling, the size of the initial analysis and assessment sets are specified. The first iteration of resampling uses these sizes, starting from the beginning of the series. The second iteration uses the same data sizes but shifts over by a set number of samples.

To illustrate, a training set of 15 samples was resampled with an analysis size of 8 samples and an assessment set size of 3 samples. The second iteration discards the first training set sample and both data sets shift forward by one. This configuration results in five resamples, as shown in Figure 10-8.

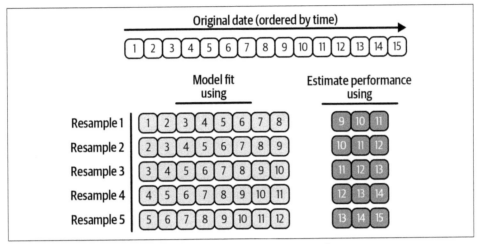

Figure 10-8. Data usage for rolling forecasting origin resampling.

There are two different configurations of this method:

- The analysis set can cumulatively grow (as opposed to remaining the same size). After the first initial analysis set, new samples can accrue without discarding the earlier data.

- The resamples need not increment by one. For example, for large data sets, the incremental block could be a week or month instead of a day.

For a year's worth of data, suppose that six sets of 30-day blocks define the analysis set. For assessment sets of 30 days with a 29-day skip, we can use the rsample package to specify:

```
time_slices <-
  tibble(x = 1:365) %>%
  rolling_origin(initial = 6 * 30, assess = 30, skip = 29, cumulative = FALSE)

data_range <- function(x) {
  summarize(x, first = min(x), last = max(x))
}

map_dfr(time_slices$splits, ~    analysis(.x) %>% data_range())
#> # A tibble: 6 × 2
#>    first  last
#>    <int> <int>
#> 1      1   180
#> 2     31   210
#> 3     61   240
```

```
#> 4    91   270
#> 5   121   300
#> 6   151   330
map_dfr(time_slices$splits, ~ assessment(.x) %>% data_range())
#> # A tibble: 6 × 2
#>   first last
#>   <int> <int>
#> 1   181   210
#> 2   211   240
#> 3   241   270
#> 4   271   300
#> 5   301   330
#> 6   331   360
```

Estimating Performance

Any of the resampling methods discussed in this chapter can be used to evaluate the modeling process (including preprocessing, model fitting, etc.). These methods are effective because different groups of data are used to train the model and assess the model. To reiterate, the process to use resampling is:

1. During resampling, the analysis set is used to preprocess the data, apply the preprocessing to itself, and use these processed data to fit the model.

2. The preprocessing statistics produced by the analysis set are applied to the assessment set. The predictions from the assessment set estimate performance on new data.

This sequence repeats for every resample. If there are B resamples, there are B replicates of each of the performance metrics. The final resampling estimate is the average of these B statistics. If $B = 1$, as with a validation set, the individual statistics represent overall performance.

Let's reconsider the previous random forest model contained in the rf_wflow object. The fit_resamples() function is analogous to fit(), but instead of having a data argument, fit_resamples() has resamples, which expects an rset object like the ones shown in this chapter. The possible interfaces to the function are:

```
model_spec %>% fit_resamples(formula, resamples, ...)
model_spec %>% fit_resamples(recipe,  resamples, ...)
workflow   %>% fit_resamples(         resamples, ...)
```

There are a number of other optional arguments, such as:

metrics

A metric set of performance statistics to compute. By default, regression models use RMSE and R^2 while classification models compute the area under the ROC curve and overall accuracy. Note that this choice also defines what predictions are produced during the evaluation of the model. For classification, if only accuracy is requested, class probability estimates are not generated for the assessment set (since they are not needed).

control

A list created by control_resamples() with various options. The control arguments include:

verbose

A logical for printing logging.

extract

A function for retaining objects from each model iteration (discussed later in this chapter).

save_pred

A logical for saving the assessment set predictions.

For our example, let's save the predictions in order to visualize the model fit and residuals:

```
keep_pred <- control_resamples(save_pred = TRUE, save_workflow = TRUE)

set.seed(1003)
rf_res <-
  rf_wflow %>%
  fit_resamples(resamples = ames_folds, control = keep_pred)
rf_res
#> # Resampling results
#> # 10-fold cross-validation
#> # A tibble: 10 × 5
#>    splits            id     .metrics        .notes           .predictions
#>    <list>            <chr>  <list>          <list>           <list>
#> 1 <split [2107/235]> Fold01 <tibble [2 × 4]> <tibble [0 × 3]> <tibble [235 × 4]>
#> 2 <split [2107/235]> Fold02 <tibble [2 × 4]> <tibble [0 × 3]> <tibble [235 × 4]>
#> 3 <split [2108/234]> Fold03 <tibble [2 × 4]> <tibble [0 × 3]> <tibble [234 × 4]>
#> 4 <split [2108/234]> Fold04 <tibble [2 × 4]> <tibble [0 × 3]> <tibble [234 × 4]>
#> 5 <split [2108/234]> Fold05 <tibble [2 × 4]> <tibble [0 × 3]> <tibble [234 × 4]>
#> 6 <split [2108/234]> Fold06 <tibble [2 × 4]> <tibble [0 × 3]> <tibble [234 × 4]>
#> # … with 4 more rows
```

The return value is a tibble similar to the input resamples, along with some extra columns:

.metrics

This is a list column of tibbles containing the assessment set performance statistics.

.notes

This is another list column of tibbles cataloging any warnings or errors generated during resampling. Note that these errors will not stop subsequent execution of resampling.

.predictions

Present when save_pred = TRUE, this list column contains tibbles with the out-of-sample predictions.

While these list columns may look daunting, they can be easily reconfigured using tidyr or with convenience functions that tidymodels provides. For example, to return the performance metrics in a more usable format:

```
collect_metrics(rf_res)
#> # A tibble: 2 × 6
#>    .metric .estimator   mean     n std_err .config
#>    <chr>   <chr>       <dbl> <int>   <dbl> <chr>
#> 1 rmse    standard    0.0721    10 0.00305 Preprocessor1_Model1
#> 2 rsq     standard    0.831     10 0.0108  Preprocessor1_Model1
```

These are the resampling estimates averaged over the individual replicates. To get the metrics for each resample, use the option summarize = FALSE.

Notice how much more realistic the performance estimates are than the resubstitution estimates from earlier in the chapter!

To obtain the assessment set predictions:

```
assess_res <- collect_predictions(rf_res)
assess_res
#> # A tibble: 2,342 × 5
#>    id      .pred  .row Sale_Price .config
#>    <chr>   <dbl> <int>      <dbl> <chr>
#> 1 Fold01   5.10    10       5.09 Preprocessor1_Model1
#> 2 Fold01   4.92    27       4.90 Preprocessor1_Model1
#> 3 Fold01   5.21    47       5.08 Preprocessor1_Model1
#> 4 Fold01   5.13    52       5.10 Preprocessor1_Model1
#> 5 Fold01   5.13    59       5.10 Preprocessor1_Model1
#> 6 Fold01   5.13    63       5.11 Preprocessor1_Model1
#> # … with 2,336 more rows
```

The prediction column names follow the conventions discussed for parsnip models in Chapter 6, for consistency and ease of use. The observed outcome column always uses the original column name from the source data. The .row column is an integer that matches the row of the original training set so that these results can be properly arranged and joined with the original data.

 For some resampling methods, such as the bootstrap or repeated cross-validation, there will be multiple predictions per row of the original training set. To obtain summarized values (averages of the replicate predictions) use collect_predictions(object, summarize = TRUE).

Since this analysis used 10-fold cross-validation, there is one unique prediction for each training set sample. These data can generate helpful plots of the model to understand where it potentially failed. For example, Figure 10-9 compares the observed and held-out predicted values (analogous to Figure 9-2):

```
assess_res %>%
  ggplot(aes(x = Sale_Price, y = .pred)) +
  geom_point(alpha = .15) +
  geom_abline(color = "red") +
  coord_obs_pred() +
  ylab("Predicted")
```

There are two houses in the training set with a low observed sale price that are significantly overpredicted by the model. Which houses are these? Let's find out from the assess_res result:

```
over_predicted <-
  assess_res %>%
  mutate(residual = Sale_Price - .pred) %>%
  arrange(desc(abs(residual))) %>%
  slice(1:2)
over_predicted
#> # A tibble: 2 × 6
#>   id     .pred .row Sale_Price .config             residual
#>   <chr>  <dbl> <int>     <dbl> <chr>                  <dbl>
#> 1 Fold09  4.97    32      4.11 Preprocessor1_Model1  -0.858
#> 2 Fold08  4.93   317      4.12 Preprocessor1_Model1  -0.815

ames_train %>%
  slice(over_predicted$.row) %>%
  select(Gr_Liv_Area, Neighborhood, Year_Built, Bedroom_AbvGr, Full_Bath)
#> # A tibble: 2 × 5
#>   Gr_Liv_Area Neighborhood         Year_Built Bedroom_AbvGr Full_Bath
#>         <int> <fct>                     <int>         <int>     <int>
#> 1         832 Old_Town                   1923             2         1
#> 2         733 Iowa_DOT_and_Rail_Road     1952             2         1
```

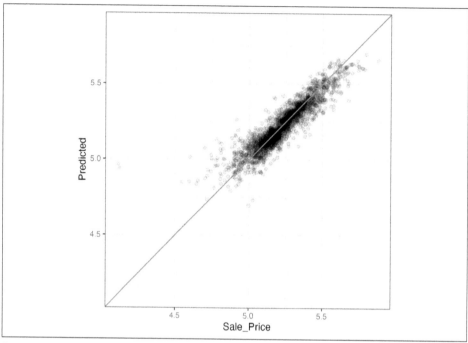

Figure 10-9. Out-of-sample observed versus predicted values for an Ames regression model, using log-10 units on both axes.

Identifying examples like these with especially poor performance can help us follow up and investigate why these specific predictions are so poor.

Let's move back to the homes overall. How can we use a validation set instead of cross-validation? From our previous rsample object:

```
val_res <- rf_wflow %>% fit_resamples(resamples = val_set)
val_res
#> # Resampling results
#> # Validation Set Split (0.75/0.25)
#> # A tibble: 1 × 4
#>   splits            id         .metrics        .notes
#>   <list>            <chr>      <list>          <list>
#> 1 <split [1756/586]> validation <tibble [2 × 4]> <tibble [0 × 3]>

collect_metrics(val_res)
#> # A tibble: 2 × 6
#>   .metric .estimator  mean     n std_err .config
#>   <chr>   <chr>      <dbl> <int>  <dbl> <chr>
#> 1 rmse    standard  0.0695     1     NA Preprocessor1_Model1
#> 2 rsq     standard  0.843      1     NA Preprocessor1_Model1
```

These results are also much closer to the test set results than the resubstitution estimates of performance.

In these analyses, the resampling results are very close to the test set results. The two types of estimates tend to be well correlated. However, this could be from random chance. A seed value of 55 fixed the random numbers before creating the resamples. Try changing this value and rerunning the analyses to investigate whether the resampled estimates match the test set results as well.

Parallel Processing

The models created during resampling are independent of one another. Computations of this kind are sometimes called "embarrassingly parallel"; each model could be fit simultaneously without issues.[9] The tune package uses the foreach (*https://oreil.ly/o821g*) package to facilitate parallel computations. These computations could be split across processors on the same computer or across different computers, depending on the chosen technology.

For computations conducted on a single computer, the number of possible worker processes is determined by the parallel package:

```
# The number of physical cores in the hardware:
parallel::detectCores(logical = FALSE)
#> [1] 10

# The number of possible independent processes that can
# be simultaneously used:
parallel::detectCores(logical = TRUE)
#> [1] 20
```

The difference between these two values is related to the computer's processor. For example, most Intel processors use hyperthreading, which creates two virtual cores for each physical core. While these extra resources can improve performance, most of the speed-ups produced by parallel processing occur when processing uses fewer than the number of physical cores.

For fit_resamples() and other functions in tune, parallel processing occurs when the user registers a parallel backend package. These R packages define how to execute parallel processing. On Unix and macOS operating systems, one method of splitting computations is by forking threads. To enable this, load the doMC package and register the number of parallel cores with foreach:

```
# Unix and macOS only
library(doMC)
registerDoMC(cores = 2)

# Now run fit_resamples()...
```

---

9 Schmidberger et al. (2009) gives a technical overview of these technologies.

This instructs `fit_resamples()` to run half of the computations on each of two cores. To reset the computations to sequential processing:

```
registerDoSEQ()
```

Alternatively, a different approach to parallelizing computations uses network sockets. The doParallel package enables this method (usable by all operating systems):

```
# All operating systems
library(doParallel)

# Create a cluster object and then register:
cl <- makePSOCKcluster(2)
registerDoParallel(cl)

# Now run fit_resamples()`...

stopCluster(cl)
```

Another R package that facilitates parallel processing is the future (*https://oreil.ly/8LLjC*) package. Like foreach, it provides a framework for parallelism. The future package is used in conjunction with foreach via the doFuture package.

> The R packages with parallel backends for foreach start with the prefix "do".

Parallel processing with the tune package tends to provide linear speed-ups for the first few cores. This means that, with two cores, the computations are twice as fast. Depending on the data and model type, the linear speed-up deteriorates after four to five cores. Using more cores will still reduce the time it takes to complete the task; there are just diminishing returns for the additional cores.

Let's wrap up with one final note about parallelism. For each of these technologies, the memory requirements multiply for each additional core used. For example, if the current data set is 2 GB in memory and three cores are used, the total memory requirement is 8 GB (2 for each worker process plus the original). Using too many cores might cause the computations (and the computer) to slow considerably.

Saving the Resampled Objects

The models created during resampling are not retained. These models are trained for the purpose of evaluating performance, and we typically do not need them after we have computed performance statistics. If a particular modeling approach does turn out to be the best option for our data set, then the best choice is to fit again to the whole training set so the model parameters can be estimated with more data.

While these models created during resampling are not preserved, there is a method for keeping them or some of their components. The `extract` option of `control_resamples()` specifies a function that takes a single argument; we'll use x. When executed, x results in a fitted workflow object, regardless of whether you provided `fit_resamples()` with a workflow. Recall that the workflows package has functions that can pull the different components of the objects (e.g., the model, recipe, etc.).

Let's fit a linear regression model using the recipe we developed in Chapter 8:

```
ames_rec <-
  recipe(Sale_Price ~ Neighborhood + Gr_Liv_Area + Year_Built + Bldg_Type +
           Latitude + Longitude, data = ames_train) %>%
  step_other(Neighborhood, threshold = 0.01) %>%
  step_dummy(all_nominal_predictors()) %>%
  step_interact( ~ Gr_Liv_Area:starts_with("Bldg_Type_") ) %>%
  step_ns(Latitude, Longitude, deg_free = 20)

lm_wflow <-
  workflow() %>%
  add_recipe(ames_rec) %>%
  add_model(linear_reg() %>% set_engine("lm"))

lm_fit <- lm_wflow %>% fit(data = ames_train)

# Select the recipe:
extract_recipe(lm_fit, estimated = TRUE)
#> Recipe
#>
#> Inputs:
#>
#>        role #variables
#>     outcome          1
#>   predictor          6
#>
#> Training data contained 2342 data points and no missing data.
#>
#> Operations:
#>
#> Collapsing factor levels for Neighborhood [trained]
#> Dummy variables from Neighborhood, Bldg_Type [trained]
#> Interactions with Gr_Liv_Area:(Bldg_Type_TwoFmCon + Bldg_Type_Duplex + B... [trained]
#> Natural splines on Latitude, Longitude [trained]
```

We can save the linear model coefficients for a fitted model object from a workflow:

```
get_model <- function(x) {
  extract_fit_parsnip(x) %>% tidy()
}

# Test it using:
# get_model(lm_fit)
```

Now let's apply this function to the ten resampled fits. The results of the extraction function are wrapped in a list object and returned in a tibble:

```
ctrl <- control_resamples(extract = get_model)

lm_res <- lm_wflow %>% fit_resamples(resamples = ames_folds, control = ctrl)
lm_res
#> # Resampling results
#> # 10-fold cross-validation
#> # A tibble: 10 × 5
#>    splits             id     .metrics         .notes           .extracts
#>    <list>             <chr>  <list>           <list>           <list>
#> 1 <split [2107/235]> Fold01 <tibble [2 × 4]> <tibble [0 × 3]> <tibble [1 × 2]>
#> 2 <split [2107/235]> Fold02 <tibble [2 × 4]> <tibble [0 × 3]> <tibble [1 × 2]>
#> 3 <split [2108/234]> Fold03 <tibble [2 × 4]> <tibble [0 × 3]> <tibble [1 × 2]>
#> 4 <split [2108/234]> Fold04 <tibble [2 × 4]> <tibble [0 × 3]> <tibble [1 × 2]>
#> 5 <split [2108/234]> Fold05 <tibble [2 × 4]> <tibble [0 × 3]> <tibble [1 × 2]>
#> 6 <split [2108/234]> Fold06 <tibble [2 × 4]> <tibble [0 × 3]> <tibble [1 × 2]>
#> # … with 4 more rows
```

Now there is a .extracts column with nested tibbles. What do these contain? Let's find out by subsetting:

```
lm_res$.extracts[[1]]
#> # A tibble: 1 × 2
#>   .extracts         .config
#>   <list>            <chr>
#> 1 <tibble [73 × 5]> Preprocessor1_Model1

# To get the results
lm_res$.extracts[[1]][[1]]
#> [[1]]
#> # A tibble: 73 × 5
#>   term                         estimate std.error statistic  p.value
#>   <chr>                           <dbl>     <dbl>     <dbl>    <dbl>
#> 1 (Intercept)                   1.48     0.320      4.62   4.11e-  6
#> 2 Gr_Liv_Area                   0.000158 0.00000476 33.2   9.72e-194
#> 3 Year_Built                    0.00180  0.000149   12.1   1.57e- 32
#> 4 Neighborhood_College_Creek   -0.00163  0.0373    -0.0438 9.65e-  1
#> 5 Neighborhood_Old_Town        -0.0757   0.0138    -5.47   4.92e-  8
#> 6 Neighborhood_Edwards         -0.109    0.0310    -3.53   4.21e-  4
#> # … with 67 more rows
```

This might appear to be a convoluted method for saving the model results. However, extract is flexible and does not assume that the user will only save a single tibble per resample. For example, the tidy() method might be run on the recipe as well as the model. In this case, a list of two tibbles will be returned.

For our more simple example, all of the results can be flattened and collected using:

```
all_coef <- map_dfr(lm_res$.extracts, ~ .x[[1]][[1]])
# Show the replicates for a single predictor:
filter(all_coef, term == "Year_Built")
#> # A tibble: 10 × 5
#>   term         estimate std.error statistic  p.value
#>   <chr>           <dbl>     <dbl>     <dbl>    <dbl>
#> 1 Year_Built    0.00180  0.000149     12.1 1.57e-32
#> 2 Year_Built    0.00180  0.000151     12.0 6.45e-32
#> 3 Year_Built    0.00185  0.000150     12.3 1.00e-33
#> 4 Year_Built    0.00183  0.000147     12.5 1.90e-34
#> 5 Year_Built    0.00184  0.000150     12.2 2.47e-33
```

```
#> 6 Year_Built  0.00180  0.000150     12.0 3.35e-32
#> # … with 4 more rows
```

Chapters 13 and 14 discuss a suite of functions for tuning models. Their interfaces are similar to `fit_resamples()`, and many of the features described here apply to those functions.

Chapter Summary

This chapter describes one of the fundamental tools of data analysis, the ability to measure the performance and variation in model results. Resampling enables us to determine how well the model works without using the test set.

An important function from the tune package, called `fit_resamples()`, was introduced. The interface for this function is also used in future chapters that describe model tuning tools.

The data analysis code, so far, for the Ames data is:

```
library(tidymodels)
data(ames)
ames <- mutate(ames, Sale_Price = log10(Sale_Price))

set.seed(502)
ames_split <- initial_split(ames, prop = 0.80, strata = Sale_Price)
ames_train <- training(ames_split)
ames_test  <- testing(ames_split)

ames_rec <-
  recipe(Sale_Price ~ Neighborhood + Gr_Liv_Area + Year_Built + Bldg_Type +
           Latitude + Longitude, data = ames_train) %>%
  step_log(Gr_Liv_Area, base = 10) %>%
  step_other(Neighborhood, threshold = 0.01) %>%
  step_dummy(all_nominal_predictors()) %>%
  step_interact( ~ Gr_Liv_Area:starts_with("Bldg_Type_") ) %>%
  step_ns(Latitude, Longitude, deg_free = 20)

lm_model <- linear_reg() %>% set_engine("lm")

lm_wflow <-
  workflow() %>%
  add_model(lm_model) %>%
  add_recipe(ames_rec)

lm_fit <- fit(lm_wflow, ames_train)

rf_model <-
  rand_forest(trees = 1000) %>%
  set_engine("ranger") %>%
  set_mode("regression")

rf_wflow <-
  workflow() %>%
  add_formula(
    Sale_Price ~ Neighborhood + Gr_Liv_Area + Year_Built + Bldg_Type +
```

```
      Latitude + Longitude) %>%
  add_model(rf_model)

set.seed(1001)
ames_folds <- vfold_cv(ames_train, v = 10)

keep_pred <- control_resamples(save_pred = TRUE, save_workflow = TRUE)

set.seed(1003)
rf_res <- rf_wflow %>% fit_resamples(resamples = ames_folds, control = keep_pred)
```

## CHAPTER 11

Comparing Models with Resampling

Once we create two or more models, the next step is to compare them to understand which one is best. In some cases, comparisons might be *within-model*, where the same model might be evaluated with different features or preprocessing methods. Alternatively, *between-model* comparisons, such as when we compared linear regression and random forest models in Chapter 10, are the more common scenario.

In either case, the result is a collection of resampled summary statistics (e.g., RMSE, accuracy, etc.) for each model. In this chapter, we'll first demonstrate how workflow sets can be used to fit multiple models. Then, we'll discuss important aspects of resampling statistics. Finally, we'll look at how to formally compare models (using either hypothesis testing or a Bayesian approach).

Creating Multiple Models with Workflow Sets

In Chapter 7 we described the idea of a workflow set where different preprocessors and/or models can be combinatorially generated. In Chapter 10, we used a recipe for the Ames data that included an interaction term as well as spline functions for longitude and latitude. To demonstrate more with workflow sets, let's create three different linear models that add these preprocessing steps incrementally; we can test whether these additional terms improve the model results. We'll create three recipes then combine them into a workflow set:

```
library(tidymodels)
tidymodels_prefer()

basic_rec <-
  recipe(Sale_Price ~ Neighborhood + Gr_Liv_Area + Year_Built + Bldg_Type +
           Latitude + Longitude, data = ames_train) %>%
  step_log(Gr_Liv_Area, base = 10) %>%
  step_other(Neighborhood, threshold = 0.01) %>%
  step_dummy(all_nominal_predictors())
```

```
interaction_rec <-
  basic_rec %>%
  step_interact( ~ Gr_Liv_Area:starts_with("Bldg_Type_") )

spline_rec <-
  interaction_rec %>%
  step_ns(Latitude, Longitude, deg_free = 50)

preproc <-
  list(basic = basic_rec,
       interact = interaction_rec,
       splines = spline_rec
  )

lm_models <- workflow_set(preproc, list(lm = linear_reg()), cross = FALSE)
lm_models
#> # A workflow set/tibble: 3 × 4
#>   wflow_id    info           option    result
#>   <chr>       <list>         <list>    <list>
#> 1 basic_lm    <tibble [1 × 4]> <opts[0]> <list [0]>
#> 2 interact_lm <tibble [1 × 4]> <opts[0]> <list [0]>
#> 3 splines_lm  <tibble [1 × 4]> <opts[0]> <list [0]>
```

We'd like to resample each of these models in turn. To do so, we will use a purrr-like function called workflow_map(). This function takes an initial argument of the function to apply to the workflows, followed by options to that function. We also set a verbose argument that will print the progress as well as a seed argument that makes sure that each model uses the same random number stream as the others:

```
lm_models <-
  lm_models %>%
  workflow_map("fit_resamples",
               # Options to `workflow_map()`:
               seed = 1101, verbose = TRUE,
               # Options to `fit_resamples()`:
               resamples = ames_folds, control = keep_pred)
#> i 1 of 3 resampling: basic_lm
#> ✓ 1 of 3 resampling: basic_lm (766ms)
#> i 2 of 3 resampling: interact_lm
#> ✓ 2 of 3 resampling: interact_lm (825ms)
#> i 3 of 3 resampling: splines_lm
#> ✓ 3 of 3 resampling: splines_lm (920ms)
lm_models
#> # A workflow set/tibble: 3 × 4
#>   wflow_id    info           option    result
#>   <chr>       <list>         <list>    <list>
#> 1 basic_lm    <tibble [1 × 4]> <opts[2]> <rsmp[+]>
#> 2 interact_lm <tibble [1 × 4]> <opts[2]> <rsmp[+]>
#> 3 splines_lm  <tibble [1 × 4]> <opts[2]> <rsmp[+]>
```

Notice that the `option` and `result` columns are now populated. The former includes the options to `fit_resamples()` that were given (for reproducibility), and the latter column contains the results produced by `fit_resamples()`.

There are a few convenience functions for workflow sets, including `collect _metrics()` to collate the performance statistics. We can `filter()` to any specific metric we are interested in:

```
collect_metrics(lm_models) %>%
  filter(.metric == "rmse")
#> # A tibble: 3 × 9
#>   wflow_id    .config         preproc model .metric .estimator   mean     n std_err
#>   <chr>       <chr>           <chr>   <chr> <chr>   <chr>       <dbl> <int>   <dbl>
#> 1 basic_lm    Preprocessor1_M… recipe line… rmse    standard   0.0803    10 0.00264
#> 2 interact_lm Preprocessor1_M… recipe line… rmse    standard   0.0799    10 0.00272
#> 3 splines_lm  Preprocessor1_M… recipe line… rmse    standard   0.0785    10 0.00282
```

What about the random forest model from the previous chapter? We can add it to the set by first converting it to its own workflow set then binding rows. This requires that, when the model was resampled, the `save_workflow = TRUE` option was set in the control function:

```
four_models <-
  as_workflow_set(random_forest = rf_res) %>%
  bind_rows(lm_models)
four_models
#> # A workflow set/tibble: 4 × 4
#>   wflow_id      info             option      result
#>   <chr>         <list>           <list>      <list>
#> 1 random_forest <tibble [1 × 4]> <opts[0]> <rsmp[+]>
#> 2 basic_lm      <tibble [1 × 4]> <opts[2]> <rsmp[+]>
#> 3 interact_lm   <tibble [1 × 4]> <opts[2]> <rsmp[+]>
#> 4 splines_lm    <tibble [1 × 4]> <opts[2]> <rsmp[+]>
```

The `autoplot()` method, with output in Figure 11-1, shows confidence intervals for each model in order of best to worst. In this chapter, we'll focus on the coefficient of determination (a.k.a. R^2) and use `metric = "rsq"` in the call to set up our plot:

```
library(ggrepel)
autoplot(four_models, metric = "rsq") +
  geom_text_repel(aes(label = wflow_id), nudge_x = 1/8, nudge_y = 1/100) +
  theme(legend.position = "none")
```

From this plot of R^2 confidence intervals, we can see that the random forest method is doing the best job and there are minor improvements in the linear models as we add more recipe steps.

Now that we have 10 resampled performance estimates for each of the four models, these summary statistics can be used to make between-model comparisons.

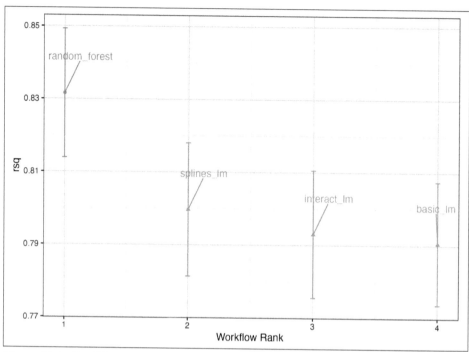

Figure 11-1. Confidence intervals for the coefficient of determination using four different models.

Comparing Resampled Performance Statistics

Considering the preceding results for the three linear models, it appears that the additional terms do not profoundly improve the mean RMSE or R^2 statistics for the linear models. The difference is small, but it might be larger than the experimental noise in the system, i.e., considered statistically significant. We can formally test the hypothesis that the additional terms increase R^2.

> Before making between-model comparisons, it is important for us to discuss the within-resample correlation for resampling statistics. Each model was measured with the same cross-validation folds, and results for the same resample tend to be similar.

In other words, there are some resamples where performance across models tends to be low and others where it tends to be high. In statistics, this is called a *resample-to-resample* component of variation.

To illustrate, let's gather the individual resampling statistics for the linear models and the random forest. We will focus on the R^2 statistic for each model, which measures correlation between the observed and predicted sale prices for each house. Let's `filter()` to keep only the R^2 metrics, reshape the results, and compute how the metrics are correlated with each other:

```
rsq_indiv_estimates <-
  collect_metrics(four_models, summarize = FALSE) %>%
  filter(.metric == "rsq")

rsq_wider <-
  rsq_indiv_estimates %>%
  select(wflow_id, .estimate, id) %>%
  pivot_wider(id_cols = "id", names_from = "wflow_id", values_from = ".estimate")

corrr::correlate(rsq_wider %>% select(-id), quiet = TRUE)
#> # A tibble: 4 × 5
#>   term          random_forest basic_lm interact_lm splines_lm
#>   <chr>                 <dbl>    <dbl>       <dbl>      <dbl>
#> 1 random_forest         NA        0.887       0.888      0.889
#> 2 basic_lm              0.887    NA           0.993      0.997
#> 3 interact_lm           0.888     0.993      NA          0.987
#> 4 splines_lm            0.889     0.997       0.987     NA
```

These correlations are high and indicate that, across models, there are large within-resample correlations. To see this visually in Figure 11-2, the R^2 statistics are shown for each model with lines connecting the resamples:

```
rsq_indiv_estimates %>%
  mutate(wflow_id = reorder(wflow_id, .estimate)) %>%
  ggplot(aes(x = wflow_id, y = .estimate, group = id, color = id, lty = id)) +
  geom_line(alpha = .8, lwd = 1.25) +
  theme(legend.position = "none")
```

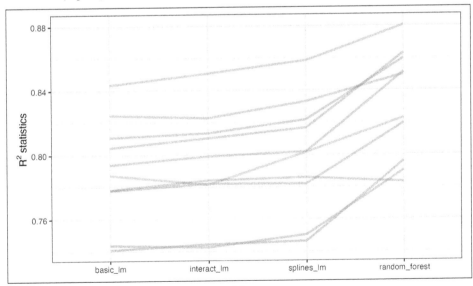

Figure 11-2. Resample statistics across models.

If the resample-to-resample effect was not real, there would not be any parallel lines. A statistical test for the correlations evaluates whether the magnitudes of these correlations are not simply noise. For the linear models:

```
rsq_wider %>%
  with( cor.test(basic_lm, splines_lm) ) %>%
  tidy() %>%
  select(estimate, starts_with("conf"))
#> # A tibble: 1 × 3
#>   estimate conf.low conf.high
#>      <dbl>    <dbl>     <dbl>
#> 1    0.997    0.987     0.999
```

The results of the correlation test (the `estimate` of the correlation and the confidence intervals) show us that the within-resample correlation appears to be real.

What effect does the extra correlation have on our analysis? Consider the variance of a difference of two variables:

$$\text{Var}\,[X - Y] = \text{Var}\,[X] + \text{Var}\,[Y] - 2\,\text{Cov}\,[X, Y]$$

The last term is the covariance between two items. If there is a significant positive covariance, then any statistical test of this difference would be critically underpowered comparing the difference in two models. In other words, ignoring the resample-to-resample effect would bias our model comparisons toward finding no differences between models.

 This characteristic of resampling statistics will come into play in the next two sections.

Before making model comparisons or looking at the resampling results, it can be helpful to define a relevant *practical effect size*. Since these analyses focus on the R^2 statistics, the practical effect size is the change in R^2 that we would consider to be a realistic difference that matters. For example, we might think that two models are not practically different if their R^2 values are within ±2%. If this were the case, differences smaller than 2% would not be deemed important even if they were statistically significant.

Practical significance is subjective; two people can have very different ideas on the threshold for importance. However, we'll show later that this consideration can be very helpful when deciding between models.

Simple Hypothesis Testing Methods

We can use simple hypothesis testing to make formal comparisons between models. Consider the familiar linear statistical model:

$$y_{ij} = \beta_0 + \beta_1 x_{i1} + ... + \beta_p x_{ip} + \epsilon_{ij}$$

This versatile model is used to create regression models as well as being the basis for the popular analysis of variance (ANOVA) technique for comparing groups. With the ANOVA model, the predictors (x_{ij}) are binary dummy variables for different groups. From this, the β parameters estimate whether two or more groups are different from one another using hypothesis testing techniques.

In our specific situation, the ANOVA can also make model comparisons. Suppose the individual resampled R^2 statistics serve as the *outcome data* (i.e., the y_{ij}) and the models as the *predictors* in the ANOVA model. A sampling of this data structure is shown in Table 11-1.

Table 11-1. Model performance statistics as a data set for analysis.

Y = rsq	model	X1	X2	X3	id
0.8108	basic_lm	0	0	0	Fold01
0.8134	interact_lm	1	0	0	Fold01
0.8615	random_forest	0	1	0	Fold01
0.8217	splines_lm	0	0	1	Fold01
0.8045	basic_lm	0	0	0	Fold02
0.8103	interact_lm	1	0	0	Fold02

The X1, X2, and X3 columns in the table are indicators for the values in the model column. Their order was defined in the same way that R would define them, alphabetically ordered by model.

For our model comparison, the specific ANOVA model is:

$$y_{ij} = \beta_0 + \beta_1 x_{i1} + \beta_2 x_{i2} + \beta_3 x_{i3} + \epsilon_{ij}$$

where

- β_0 is the estimate of the mean R^2 statistic for the basic linear models (i.e., without splines or interactions).
- β_1 is the change in mean R^2 when interactions are added to the basic linear model.

- β_2 is the change in mean R^2 between the basic linear model and the random forest model.

- β_3 is the change in mean R^2 between the basic linear model and one with interactions and splines.

From these model parameters, hypothesis tests and p-values are generated to statistically compare models, but we must contend with how to handle the resample-to-resample effect. Historically, the resample groups would be considered a *block effect* and an appropriate term was added to the model. Alternatively, the resample effect could be considered a *random effect* where these particular resamples were drawn at random from a larger population of possible resamples. However, we aren't really interested in these effects; we only want to adjust for them in the model so that the variances of the interesting differences are properly estimated.

Treating the resamples as random effects is theoretically appealing. Methods for fitting an ANOVA model with this type of random effect could include the linear mixed model (Faraway 2016) or a Bayesian hierarchical model (shown in the next section).

A simple and fast method for comparing two models at a time is to use the differences in R^2 values as the outcome data in the ANOVA model. Since the outcomes are matched by resample, the differences do not contain the resample-to-resample effect and, for this reason, the standard ANOVA model is appropriate. To illustrate, this call to lm() tests the difference between two of the linear regression models:

```
compare_lm <-
  rsq_wider %>%
  mutate(difference = splines_lm - basic_lm)

lm(difference ~ 1, data = compare_lm) %>%
  tidy(conf.int = TRUE) %>%
  select(estimate, p.value, starts_with("conf"))
#> # A tibble: 1 × 4
#>   estimate   p.value conf.low conf.high
#>      <dbl>     <dbl>    <dbl>     <dbl>
#> 1  0.00913 0.0000256  0.00650    0.0118

# Alternatively, a paired t-test could also be used:
rsq_wider %>%
  with( t.test(splines_lm, basic_lm, paired = TRUE) ) %>%
  tidy() %>%
  select(estimate, p.value, starts_with("conf"))
#> # A tibble: 1 × 4
#>   estimate   p.value conf.low conf.high
#>      <dbl>     <dbl>    <dbl>     <dbl>
#> 1  0.00913 0.0000256  0.00650    0.0118
```

We could evaluate each pair-wise difference in this way. Note that the p-value indicates a *statistically significant* signal; the collection of spline terms for longitude and latitude do appear to have an effect. However, the difference in R^2 is estimated at

0.91%. If our practical effect size were 2%, we might not consider these terms worth including in the model.

 We've briefly mentioned p-values already, but what actually are they? From Wasserstein and Lazar (2016): "Informally, a p-value is the probability under a specified statistical model that a statistical summary of the data (e.g., the sample mean difference between two compared groups) would be equal to or more extreme than its observed value."

In other words, if this analysis were repeated a large number of times under the null hypothesis of no differences, the p-value reflects how extreme our observed results would be in comparison.

Bayesian Methods

We just used hypothesis testing to formally compare models, but we can also take a more general approach to making these formal comparisons using random effects and Bayesian statistics (McElreath 2020). While the model is more complex than the ANOVA method, the interpretation is more simple and straightforward than the p-value approach. The previous ANOVA model had the form:

$$y_{ij} = \beta_0 + \beta_1 x_{i1} + \beta_2 x_{i2} + \beta_3 x_{i3} + \epsilon_{ij}$$

where the residuals ϵ_{ij} are assumed to be independent and follow a Gaussian distribution with zero mean and constant standard deviation of σ. From this assumption, statistical theory shows that the estimated regression parameters follow a multivariate Gaussian distribution and, from this, p-values and confidence intervals are derived.

A Bayesian linear model makes additional assumptions. In addition to specifying a distribution for the residuals, we require *prior distribution* specifications for the model parameters (β_j and σ). These are distributions for the parameters that the model assumes before being exposed to the observed data. For example, a simple set of prior distributions for our model might be:

$$\epsilon_{ij} \sim N(0, \sigma)$$
$$\beta_j \sim N(0, 10)$$
$$\sigma \sim \text{exponential}(1)$$

These priors set the possible/probable ranges of the model parameters and have no unknown parameters. For example, the prior on σ indicates that values must be larger than zero, are very right-skewed, and have values that are usually less than 3 or 4.

Note that the regression parameters have a pretty wide prior distribution, with a standard deviation of 10. In many cases, we might not have a strong opinion about the prior beyond it being symmetric and bell shaped. The large standard deviation implies a fairly uninformative prior; it is not overly restrictive in terms of the possible values that the parameters might take on. This allows the data to have more of an influence during parameter estimation.

Given the observed data and the prior distribution specifications, the model parameters can then be estimated. The final distributions of the model parameters are combinations of the priors and the likelihood estimates. These *posterior distributions* of the parameters are the key distributions of interest. They are a full probabilistic description of the model's estimated parameters.

A Random Intercept Model

To adapt our Bayesian ANOVA model so that the resamples are adequately modeled, we consider a *random intercept model*. Here, we assume that the resamples impact the model only by changing the intercept. Note that this constrains the resamples from having a differential impact on the regression parameters β_j; these are assumed to have the same relationship across resamples. This model equation is:

$$y_{ij} = (\beta_0 + b_i) + \beta_1 x_{i1} + \beta_2 x_{i2} + \beta_3 x_{i3} + \epsilon_{ij}$$

This is not an unreasonable model for resampled statistics which, when plotted across models as in Figure 11-2, tend to have fairly parallel effects across models (i.e., little crossover of lines).

For this model configuration, an additional assumption is made for the prior distribution of random effects. A reasonable assumption for this distribution is another symmetric distribution, such as another bell-shaped curve. Given the effective sample size of 10 in our summary statistic data, let's use a prior that is wider than a standard normal distribution. We'll use a t-distribution with a single degree of freedom (i.e., $b_i \sim t(1)$), which has heavier tails than an analogous Gaussian distribution.

The tidyposterior package has functions to fit such Bayesian models for the purpose of comparing resampled models. The main function is called perf_mod(), and it is configured to "just work" for different types of objects:

- For workflow sets, it creates an ANOVA model where the groups correspond to the workflows. Our existing models did not optimize any tuning parameters (see the next three chapters). If one of the workflows in the set had data on tuning parameters, the best tuning parameters set for each workflow is used in the Bayesian analysis. In other words, despite the presence of tuning parameters, perf_mod() focuses on making *between-workflow comparisons*.

- For objects that contain a single model that has been tuned using resampling, `perf_mod()` makes *within-model comparisons*. In this situation, the grouping variables tested in the Bayesian ANOVA model are the submodels defined by the tuning parameters.

- The `perf_mod()` function can also take a data frame produced by rsample that has columns of performance metrics associated with two or more model/workflow results. These could have been generated by nonstandard means.

From any of these types of objects, the `perf_mod()` function determines an appropriate Bayesian model and fits it with the resampling statistics. For our example, it will model the four sets of R^2 statistics associated with the workflows.

The tidyposterior package uses the Stan software (*https://mc-stan.org*) for specifying and fitting the models via the rstanarm package. The functions within that package have default priors (see `?priors` for more details). The following model uses the default priors for all parameters except for the random intercepts (which follow a *t*-distribution). The estimation process uses random numbers so the seed is set within the function call. The estimation process is iterative and replicated several times in collections called *chains*. The `iter` parameter tells the function how long to run the estimation process in each chain. When several chains are used, their results are combined (assume that this is validated by diagnostic assessments):

```
library(tidyposterior)
library(rstanarm)

# The rstanarm package creates copious amounts of output; those results
# are not shown here but are worth inspecting for potential issues. The
# option `refresh = 0` can be used to eliminate the logging.
rsq_anova <-
  perf_mod(
    four_models,
    metric = "rsq",
    prior_intercept = rstanarm::student_t(df = 1),
    chains = 4,
    iter = 5000,
    seed = 1102
  )
```

The resulting object has information on the resampling process as well as the Stan object embedded within (in an element called `stan`). We are most interested in the posterior distributions of the regression parameters. The tidyposterior package has a `tidy()` method that extracts these posterior distributions into a tibble:

```
model_post <-
  rsq_anova %>%
  # Take a random sample from the posterior distribution
  # so set the seed again to be reproducible.
  tidy(seed = 1103)

glimpse(model_post)
```

```
#> Rows: 40,000
#> Columns: 2
#> $ model     <chr> "random_forest", "random_forest", "random_forest", …
#> $ posterior <dbl> 0.8293, 0.8238, 0.8276, 0.8209, 0.8213, 0.8132, 0.8241, …
```

The four posterior distributions are visualized in Figure 11-3:

```
model_post %>%
  mutate(model = forcats::fct_inorder(model)) %>%
  ggplot(aes(x = posterior)) +
  geom_histogram(bins = 50, color = "white", fill = "blue", alpha = 0.4) +
  facet_wrap(~ model, ncol = 1)
```

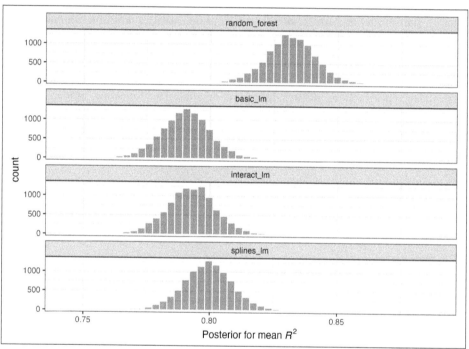

Figure 11-3. Posterior distributions for the coefficient of determination using four different models.

These histograms describe the estimated probability distributions of the mean R^2 value for each model. There is some overlap, especially for the three linear models.

There is also a basic `autoplot()` method for the model results, shown in Figure 11-4, as well as the tidied object that shows overlaid density plots:

```
autoplot(rsq_anova) +
  geom_text_repel(aes(label = workflow), nudge_x = 1/8, nudge_y = 1/100) +
  theme(legend.position = "none")
```

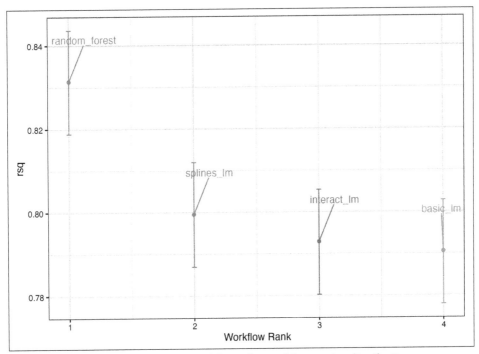

Figure 11-4. Credible intervals derived from the model posterior distributions.

One wonderful aspect of using resampling with Bayesian models is that, once we have the posteriors for the parameters, it is trivial to get the posterior distributions for combinations of the parameters. For example, to compare the two linear regression models, we are interested in the difference in means. The posterior of this difference is computed by sampling from the individual posteriors and taking the differences. The `contrast_models()` function can do this. To specify the comparisons to make, the `list_1` and `list_2` parameters take character vectors and compute the differences between the models in those lists (parameterized as `list_1 - list_2`).

We can compare two of the linear models and visualize the results in Figure 11-5:

```
rqs_diff <-
  contrast_models(rsq_anova,
                  list_1 = "splines_lm",
                  list_2 = "basic_lm",
                  seed = 1104)

rqs_diff %>%
  as_tibble() %>%
  ggplot(aes(x = difference)) +
  geom_vline(xintercept = 0, lty = 2) +
  geom_histogram(bins = 50, color = "white", fill = "red", alpha = 0.4)
```

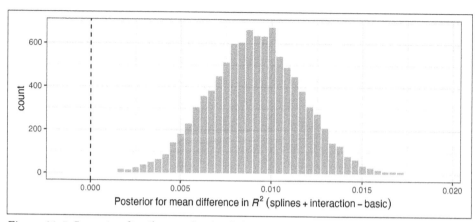

Figure 11-5. Posterior distribution for the difference in the coefficient of determination.

The posterior shows that the center of the distribution is greater than zero (indicating that the model with splines typically had larger values) but does overlap with zero to a degree. The `summary()` method for this object computes the mean of the distribution as well as credible intervals, the Bayesian analog to confidence intervals:

```
summary(rqs_diff) %>%
  select(-starts_with("pract"))
#> # A tibble: 1 × 6
#>   contrast            probability   mean  lower  upper  size
#>   <chr>                     <dbl>  <dbl>  <dbl>  <dbl> <dbl>
#> 1 splines_lm vs basic_lm        1 0.00913 0.00507 0.0131     0
```

The `probability` column reflects the proportion of the posterior that is greater than zero. This is the probability that the positive difference is real. The value is not close to zero, providing a strong case for statistical significance, i.e., the idea that statistically the actual difference is not zero.

However, the estimate of the mean difference is fairly close to zero. Recall that the practical effect size we suggested previously is 2%. With a posterior distribution, we can also compute the probability of being practically significant. In Bayesian analysis, this is a *ROPE estimate*, or Region Of Practical Equivalence (Kruschke and Liddell 2018). To estimate this, the `size` option to the summary function is used:

```
summary(rqs_diff, size = 0.02) %>%
  select(contrast, starts_with("pract"))
#> # A tibble: 1 × 4
#>   contrast            pract_neg pract_equiv pract_pos
#>   <chr>                   <dbl>       <dbl>     <dbl>
#> 1 splines_lm vs basic_lm      0           1         0
```

The `pract_equiv` column is the proportion of the posterior that is within [-`size`, `size`] (the columns `pract_neg` and `pract_pos` are the proportions that are below and above this interval). This large value indicates that, for our effect size, there is an overwhelming probability that the two models are practically the same. Even though the previous plot showed that our difference is likely nonzero, the equivalence test suggests that it is small enough to not be practically meaningful.

The same process could be used to compare the random forest model to one or both of the linear regressions that were resampled. In fact, when perf_mod() is used with a workflow set, the autoplot() method can show the pract_equiv results that compare each workflow to the current best (the random forest model, in this case):

```
autoplot(rsq_anova, type = "ROPE", size = 0.02) +
  geom_text_repel(aes(label = workflow)) +
  theme(legend.position = "none")
```

Figure 11-6 shows us that none of the linear models comes close to the random forest model when a 2% practical effect size is used.

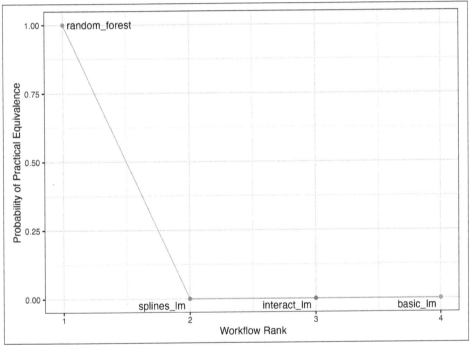

Figure 11-6. Probability of practical equivalence for an effect size of 2%.

The Effect of the Amount of Resampling

How does the number of resamples affect these types of formal Bayesian comparisons? More resamples increases the precision of the overall resampling estimate; that precision propagates to this type of analysis. For illustration, more resamples were added using repeated cross-validation. How did the posterior distribution change?

Figure 11-7 shows the 90% credible intervals with up to 100 resamples (generated from 10 repeats of 10-fold cross-validation):[1]

```
ggplot(intervals,
       aes(x = resamples, y = mean)) +
  geom_path() +
  geom_ribbon(aes(ymin = lower, ymax = upper), fill = "red", alpha = .1) +
  labs(x = "Number of Resamples (repeated 10-fold cross-validation)")
```

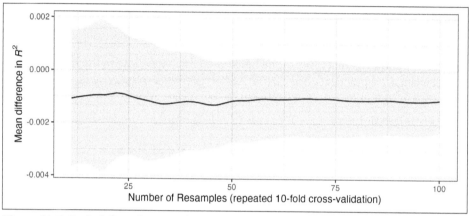

Figure 11-7. Probability of practical equivalence to the random forest model.

The width of the intervals decreases as more resamples are added. Clearly, going from 10 resamples to 30 has a larger impact than going from 80 to 100. There are diminishing returns for using a "large" number of resamples ("large" will be different for different data sets).

Chapter Summary

This chapter described formal statistical methods for testing differences in performance between models. We demonstrated the within-resample effect, where results for the same resample tend to be similar; this aspect of resampled summary statistics requires appropriate analysis in order for valid model comparisons. Further, although statistical significance and practical significance are both important concepts for model comparisons, they are different.

---

1 The code to generate intervals is available at GitHub (*https://oreil.ly/CmvNU*).

## CHAPTER 12

Model Tuning and the Dangers of Overfitting

In order to use a model for prediction, the parameters for that model must be estimated. Some of these parameters can be estimated directly from the training data, but other parameters, called *tuning parameters* or *hyperparameters*, must be specified ahead of time and can't be directly found from training data. These are unknown structural values or other kinds of values that have significant impact on the model but cannot be directly estimated from the data. This chapter will provide examples of tuning parameters and show how we use the tidymodels functions to create and handle tuning parameters. We'll also demonstrate how poor choices of these values lead to overfitting and introduce several tactics for finding optimal tuning parameters values. Chapters 13 and 14 go into more detail on specific optimization methods for tuning.

Model Parameters

In ordinary linear regression, there are two parameters β_0 and β_1 of the model:

$$y_i = \beta_0 + \beta_1 x_i + \epsilon_i$$

When we have the outcome (y) and predictor (x) data, we can estimate the two parameters β_0 and β_1:

$$\widehat{\beta}_1 = \frac{\Sigma_i (y_i - \bar{y})(x_i - \bar{x})}{\Sigma_i (x_i - \bar{x})^2}$$

and

$$\widehat{\beta}_0 = \bar{y} - \widehat{\beta}_1 \bar{x}.$$

We can directly estimate these values from the data for this example model because they are analytically tractable; if we have the data, then we can estimate these model parameters.

 There are many situations where a model has parameters that *can't* be directly estimated from the data.

For the KNN model, the prediction equation for a new value x_0 is:

$$\widehat{y} = \frac{1}{K} \sum_{\ell=1}^{K} x_\ell^*$$

where K is the number of neighbors and the x_ℓ^* are the K closest values to x_0 in the training set. The model itself is not defined by a model equation; the previous prediction equation instead defines it. This characteristic, along with the possible intractability of the distance measure, makes it impossible to create a set of equations that can be solved for K (iteratively or otherwise). The number of neighbors has a profound impact on the model; it governs the flexibility of the class boundary. For small values of K, the boundary is very elaborate, while for large values, it might be quite smooth.

The number of nearest neighbors is a good example of a tuning parameter or hyperparameter that cannot be directly estimated from the data.

Tuning Parameters for Different Types of Models

There are many examples of tuning parameters or hyperparameters in different statistical and machine learning models:

- Boosting is an ensemble method that combines a series of base models, each of which is created sequentially and depends on the previous models. The number of boosting iterations is an important tuning parameter that usually requires optimization.

- In the classic single-layer artificial neural network (a.k.a. the multilayer perceptron), the predictors are combined using two or more hidden units. The hidden units are linear combinations of the predictors that are captured in an *activation function* (typically a nonlinear function, such as a sigmoid). The hidden units are then connected to the outcome units; one outcome unit is used for regression models, and multiple outcome units are required for classification. The number of hidden units and the type of activation function are important structural tuning parameters.

- Modern gradient descent methods are improved by finding the right optimization parameters. Examples of such hyperparameters are learning rates, momentum, and the number of optimization iterations/epochs (Goodfellow, Bengio, and Courville 2016). Neural networks and some ensemble models use gradient descent to estimate the model parameters. While the tuning parameters associated with gradient descent are not structural parameters, they often require tuning.

In some cases, preprocessing techniques require tuning:

- In principal component analysis, or its supervised cousin called partial least squares, the predictors are replaced with new, artificial features that have better properties related to collinearity. The number of extracted components can be tuned.

- Imputation methods estimate missing predictor values using the complete values of one or more predictors. One effective imputation tool uses K-nearest neighbors of the complete columns to predict the missing value. The number of neighbors modulates the amount of averaging and can be tuned.

Some classical statistical models also have structural parameters:

- In binary regression, the logit link is commonly used (i.e., logistic regression). Other link functions, such as the probit and complementary log-log, are also available (Dobson 1999). This example is described in more detail in the next section.

- Non-Bayesian longitudinal and repeated measures models require a specification for the covariance or correlation structure of the data. Options include compound symmetric (a.k.a. exchangeable), autoregressive, Toeplitz, and others (Littell, Pendergast, and Natarajan 2000).

A counterexample in which it is inappropriate to tune a parameter is the prior distribution required for Bayesian analysis. The prior encapsulates the analyst's belief about the distribution of a quantity before evidence or data are taken into account. For example, in Chapter 11, we used a Bayesian ANOVA model, and we were unclear about what the prior should be for the regression parameters (beyond being a symmetric distribution). We chose a t-distribution with one degree of freedom for the prior since it has heavier tails; this reflects our added uncertainty. Our prior beliefs should not be subject to optimization. Tuning parameters are typically optimized for performance whereas priors should not be tweaked to get "the right results."

 Another (perhaps more debatable) counterexample of a parameter that does *not* need to be tuned is the number of trees in a random forest or bagging model. This value should instead be chosen to be large enough to ensure numerical stability in the results; tuning it cannot improve performance as long as the value is large enough to produce reliable results. For random forests, this value is typically in the thousands while the number of trees needed for bagging is around 50 to 100.

What Do We Optimize?

How should we evaluate models when we optimize tuning parameters? It depends on the model and the purpose of the model.

For cases where the statistical properties of the tuning parameter are tractable, common statistical properties can be used as the objective function. For example, in the case of binary logistic regression, the link function can be chosen by maximizing the likelihood or information criteria. However, these statistical properties may not align with the results achieved using accuracy-oriented properties. As an example, Friedman (2001) optimized the number of trees in a boosted tree ensemble and found different results when maximizing the likelihood and accuracy:

> degrading the likelihood by overfitting actually improves misclassification error rate. Although perhaps counterintuitive, this is not a contradiction; likelihood and error rate measure different aspects of fit quality.

To demonstrate, consider the classification data shown in Figure 12-1 with two predictors, two classes, and a training set of 593 data points.

We could start by fitting a linear class boundary to these data. The most common method for doing this is to use a generalized linear model in the form of *logistic regression*. This model relates the *log odds* of a sample being Class 1 using the *logit* transformation:

$$\log\left(\frac{\pi}{1-\pi}\right) = \beta_0 + \beta_1 x_1 + \dots + \beta_p x_p$$

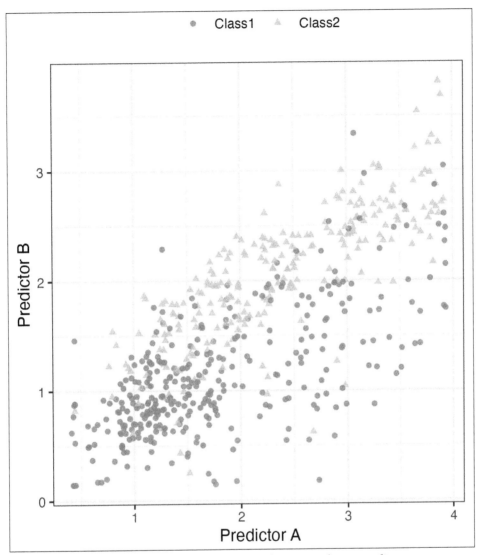

Figure 12-1. An example two-class classification data set with two predictors.

In the context of generalized linear models, the logit function is the *link function* between the outcome (π) and the predictors. There are other link functions that include the *probit* model:

$$\Phi^{-1}(\pi) = \beta_0 + \beta_1 x_1 + \ldots + \beta_p x_p$$

where Φ is the cumulative standard normal function, as well as the *complementary log-log* model:

$$\log\left(-\log\left(1-\pi\right)\right) = \beta_0 + \beta_1 x_1 + \ldots + \beta_p x_p$$

Each of these models results in linear class boundaries. Which one should we use? Since, for these data, the number of model parameters does not vary, the statistical approach is to compute the (log) likelihood for each model and determine the model with the largest value. Traditionally, the likelihood is computed using the same data that were used to estimate the parameters, not using approaches like data splitting or resampling from Chapters 5 and 10.

For a data frame `training_set`, let's create a function to compute the different models and extract the likelihood statistics for the training set (using `broom::glance()`):

```
library(tidymodels)
tidymodels_prefer()

llhood <- function(...) {
  logistic_reg() %>%
    set_engine("glm", ...) %>%
    fit(Class ~ ., data = training_set) %>%
    glance() %>%
    select(logLik)
}

bind_rows(
  llhood(),
  llhood(family = binomial(link = "probit")),
  llhood(family = binomial(link = "cloglog"))
) %>%
  mutate(link = c("logit", "probit", "c-log-log"))  %>%
  arrange(desc(logLik))
#> # A tibble: 3 × 2
#>    logLik link
#>     <dbl> <chr>
#> 1   -258. logit
#> 2   -262. probit
#> 3   -270. c-log-log
```

According to these results, the logistic model has the best statistical properties.

From the scale of the log-likelihood values, it is difficult to understand if these differences are important or negligible. One way of improving this analysis is to resample the statistics and separate the modeling data from the data used for performance estimation. With this small data set, repeated 10-fold cross-validation is a good choice for resampling. In the yardstick package, the `mn_log_loss()` function is used to estimate the negative log-likelihood, with our results shown in Figure 12-2:

```
set.seed(1201)
rs <- vfold_cv(training_set, repeats = 10)

# Return the individual resampled performance estimates:
lloss <- function(...) {
  perf_meas <- metric_set(roc_auc, mn_log_loss)

  logistic_reg() %>%
    set_engine("glm", ...) %>%
    fit_resamples(Class ~ A + B, rs, metrics = perf_meas) %>%
    collect_metrics(summarize = FALSE) %>%
    select(id, id2, .metric, .estimate)
}

resampled_res <-
  bind_rows(
    lloss()                                     %>% mutate(model = "logistic"),
    lloss(family = binomial(link = "probit"))  %>% mutate(model = "probit"),
    lloss(family = binomial(link = "cloglog")) %>% mutate(model = "c-log-log")
  ) %>%
  # Convert log-loss to log-likelihood:
  mutate(.estimate = ifelse(.metric == "mn_log_loss", -.estimate, .estimate)) %>%
  group_by(model, .metric) %>%
  summarize(
    mean = mean(.estimate, na.rm = TRUE),
    std_err = sd(.estimate, na.rm = TRUE) / sum(!is.na(.estimate)),
    .groups = "drop"
  )

resampled_res %>%
  filter(.metric == "mn_log_loss") %>%
  ggplot(aes(x = mean, y = model)) +
  geom_point() +
  geom_errorbar(aes(xmin = mean - 1.64 * std_err, xmax = mean + 1.64 * std_err),
                width = .1) +
  labs(y = NULL, x = "log-likelihood")
```

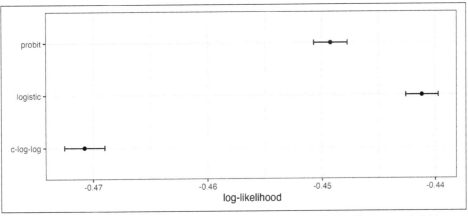

Figure 12-2. Means and approximate 90% confidence intervals for the resampled bino-mial log-likelihood with three different link functions.

The scale of these values is different than the previous values since they are computed on a smaller data set; the value produced by broom::glance() is a sum while yardstick::mn_log_loss() is an average.

These results show that there is considerable evidence that the choice of the link function matters and that the logistic model is superior.

What about a different metric? We also calculated the area under the ROC curve for each resample. These results, which reflect the discriminative ability of the models across numerous probability thresholds, show a lack of difference in Figure 12-3.

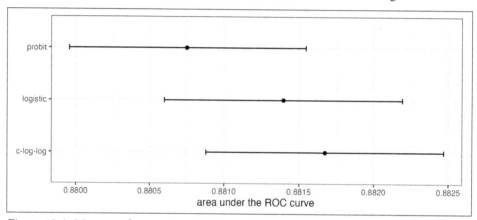

Figure 12-3. Means and approximate 90% confidence intervals for the resampled area under the ROC curve with three different link functions.

Given the overlap of the intervals, as well as the scale of the x-axis, any of these options could be used. We see this again when the class boundaries for the three models are overlaid on the test set of 198 data points in Figure 12-4.

This exercise emphasizes that different metrics might lead to different decisions about the choice of tuning parameter values. In this case, one metric appears to clearly sort the models while another metric shows no difference.

Metric optimization is thoroughly discussed by Thomas and Uminsky (2020) who explore several issues, including the gaming of metrics. They warn that:

> The unreasonable effectiveness of metric optimization in current AI approaches is a fundamental challenge to the field, and yields an inherent contradiction: solely optimizing metrics leads to far from optimal outcomes.

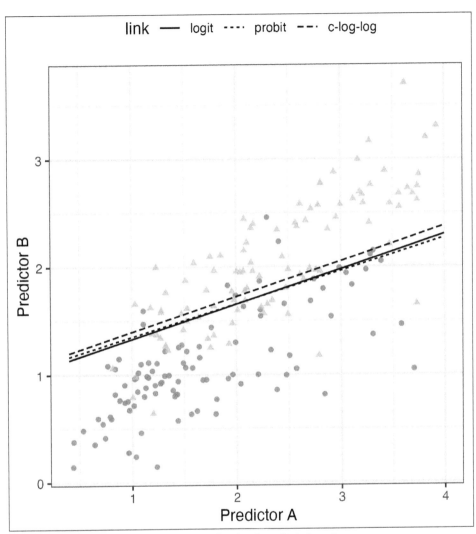

Figure 12-4. The linear class boundary fits for three link functions.

The Consequences of Poor Parameter Estimates

Many tuning parameters modulate the amount of model complexity. More complexity often implies more malleability in the patterns that a model can emulate. For example, as shown in Chapter 8, adding degrees of freedom in a spline function increases the intricacy of the prediction equation. While this is an advantage when the underlying motifs in the data are complex, it can also lead to overinterpretation of chance patterns that would not reproduce in new data. *Overfitting* is the situation where a model adapts too much to the training data; it performs well for the data used to build the model but poorly for new data.

 Since tuning model parameters can increase model complexity, poor choices can lead to overfitting.

Recall the single-layer neural network model described in the first section of this chapter. With a single hidden unit and sigmoidal activation functions, a neural network for classification is, for all intents and purposes, just logistic regression. However, as the number of hidden units increases, so does the complexity of the model. In fact, Cybenko (1989) showed that, when the network model uses sigmoidal activation units, the model is a universal function approximator as long as there are enough hidden units.

We fit neural network classification models to the same two-class data from the previous section, varying the number of hidden units. Using the area under the ROC curve as a performance metric, the effectiveness of the model on the training set increases as more hidden units are added. The network model thoroughly and meticulously learns the training set. If the model judges itself on the training set ROC value, it prefers many hidden units so that it can nearly eliminate errors.

Chapters 5 and 10 demonstrated that simply repredicting the training set is a poor approach to model evaluation. Here, the neural network very quickly begins to overinterpret patterns that it sees in the training set. Compare these three example class boundaries (developed with the training set) overlaid on training and test sets in Figure 12-5.

The single-unit model does not adapt very flexibly to the data (since it is constrained to be linear). A model with four hidden units begins to show signs of overfitting with an unrealistic boundary for values away from the data mainstream. This is caused by a single data point from the first class in the upper-right corner of the data. By 20 hidden units, the model is beginning to memorize the training set, creating small islands around those data to minimize the resubstitution error rate. These patterns do not repeat in the test set. This last panel is the best illustration of how tuning parameters that control complexity must be modulated so that the model is effective. For a 20-unit model, the training set area under the ROC curve (ROC AUC) is 0.944 but the test set value is 0.855.

This occurrence of overfitting is obvious with two predictors that we can plot. However, in general, we must use a quantitative approach for detecting overfitting.

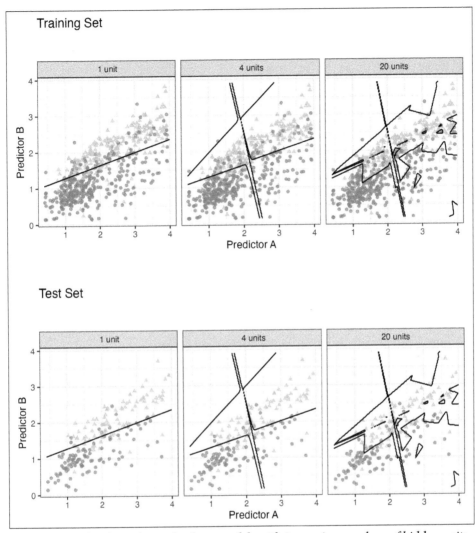

Figure 12-5. Class boundaries for three models with increasing numbers of hidden units. The boundaries are fit on the training set and shown for the training and test sets.

The solution for detecting when a model is overemphasizing the training set is using out-of-sample data.

Rather than using the test set, some form of resampling is required. This could mean an iterative approach (e.g., 10-fold cross-validation) or a single data source (e.g., a validation set).

Two General Strategies for Optimization

Tuning parameter optimization usually falls into one of two categories: grid search and iterative search.

Grid search is when we predefine a set of parameter values to evaluate. The main choices involved in grid search are how to make the grid and how many parameter combinations to evaluate. Grid search is often judged as inefficient since the number of grid points required to cover the parameter space can become unmanageable with the curse of dimensionality. There is truth to this concern, but it is most true when the process is not optimized. This is discussed more in Chapter 13.

Iterative search or sequential search is when we sequentially discover new parameter combinations based on previous results. Almost any nonlinear optimization method is appropriate, although some are more efficient than others. In some cases, an initial set of results for one or more parameter combinations is required to start the optimization process. Iterative search is discussed more in Chapter 14.

Figure 12-6 shows two panels that demonstrate these two approaches for a situation with two tuning parameters that range between zero and one. In each, a set of contours shows the true (simulated) relationship between the parameters and the outcome. The optimal results are in the upper-right corner.

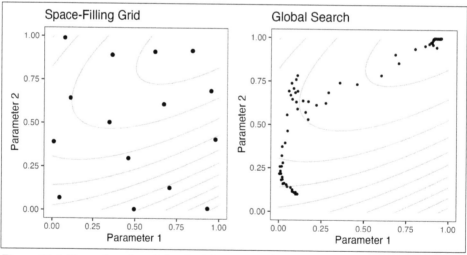

Figure 12-6. Examples of predefined grid tuning and an iterative search method. The lines represent contours of a performance metric; it is best in the upper-right side of the plot.

The left-hand panel of Figure 12-6 shows a type of grid called a space-filling design. This is a type of experimental design devised for covering the parameter space such that tuning parameter combinations are not close to one another. The results for this design do not place any points exactly at the truly optimal location. However, one point is in the general vicinity and would probably have performance metric results that are within the noise of the most optimal value.

The right-hand panel of Figure 12-6 illustrates the results of a global search method: the Nelder–Mead simplex method (Olsson and Nelson 1975). The starting point is in the lower-left part of the parameter space. The search meanders across the space until it reaches the optimum location, where it strives to come as close as possible to the numerically best value. This particular search method, while effective, is not known for its efficiency; it requires many function evaluations, especially near the optimal values. Chapter 14 discusses more efficient search algorithms.

 Hybrid strategies are also an option and can work well. After an initial grid search, a sequential optimization can start from the best grid combination.

Examples of these strategies are discussed in detail in the next two chapters. Before moving on, let's learn how to work with tuning parameter objects in tidymodels, using the dials package.

Tuning Parameters in tidymodels

We've already dealt with quite a number of arguments that correspond to tuning parameters for recipe and model specifications in previous chapters. It is possible to tune:

- The threshold for combining neighborhoods into an "other" category (with argument name `threshold`) discussed in Chapter 8
- The number of degrees of freedom in a natural spline (`deg_free`, Chapter 8)
- The number of data points required to execute a split in a tree-based model (`min_n`, Chapter 6)
- The amount of regularization in penalized models (`penalty`, Chapter 6)

For parsnip model specifications, there are two kinds of parameter arguments. *Main arguments* are those that are most often optimized for performance and are available in multiple engines. The main tuning parameters are top-level arguments to the model specification function. For example, the `rand_forest()` function has main arguments `trees`, `min_n`, and `mtry` since these are most frequently specified or optimized.

A secondary set of tuning parameters are *engine specific*. These are either infrequently optimized or are specific only to certain engines. Again using random forests as an example, the ranger package contains some arguments that are not used by other packages. One example is gain penalization, which regularizes the predictor selection in the tree induction process. This parameter can help modulate the trade-off between the number of predictors used in the ensemble and performance (Wundervald, Parnell, and Domijan 2020). The name of this argument in `ranger()` is `regularization.factor`. To specify a value via a parsnip model specification, it is added as a supplemental argument to `set_engine()`:

```
rand_forest(trees = 2000, min_n = 10) %>%          # <- main arguments
  set_engine("ranger", regularization.factor = 0.5)  # <- engine-specific
```

 The main arguments use a harmonized naming system to remove inconsistencies across engines while engine-specific arguments do not.

How can we signal to tidymodels functions which arguments should be optimized? Parameters are marked for tuning by assigning them a value of `tune()`. For the single-layer neural network used earlier in this chapter, the number of hidden units is designated for tuning using:

```
neural_net_spec <-
  mlp(hidden_units = tune()) %>%
  set_engine("keras")
```

The `tune()` function doesn't execute any particular parameter value; it only returns an expression:

```
tune()
#> tune()
```

Embedding this `tune()` value in an argument will tag the parameter for optimization. The model tuning functions shown in the next two chapters parse the model specification and/or recipe to discover the tagged parameters. These functions can automatically configure and process these parameters since they understand their characteristics (e.g., the range of possible values, etc.).

Use the `extract_parameter_set_dials()` function to enumerate the tuning parameters for an object:

```
extract_parameter_set_dials(neural_net_spec)
#> Collection of 1 parameters for tuning
#>
#>    identifier        type    object
#>    hidden_units hidden_units nparam[+]
```

The results show a value of `nparam[+]`, indicating that the number of hidden units is a numeric parameter.

There is an optional identification argument that associates a name with the parameters. This can come in handy when the same kind of parameter is being tuned in different places. For example, with the Ames housing data example from the end of Chapter 10, the recipe encoded both longitude and latitude with spline functions. If we want to tune the two spline functions to potentially have different levels of smoothness, we call `step_ns()` twice, once for each predictor. To make the parameters identifiable, the identification argument can take any character string:

```
ames_rec <-
  recipe(Sale_Price ~ Neighborhood + Gr_Liv_Area + Year_Built + Bldg_Type +
         Latitude + Longitude, data = ames_train)  %>%
  step_log(Gr_Liv_Area, base = 10) %>%
  step_other(Neighborhood, threshold = tune()) %>%
  step_dummy(all_nominal_predictors()) %>%
  step_interact( ~ Gr_Liv_Area:starts_with("Bldg_Type_") ) %>%
  step_ns(Longitude, deg_free = tune("longitude df")) %>%
  step_ns(Latitude,  deg_free = tune("latitude df"))

recipes_param <- extract_parameter_set_dials(ames_rec)
recipes_param
#> Collection of 3 parameters for tuning
#>
#>    identifier      type    object
#>     threshold threshold nparam[+]
#>  longitude df  deg_free nparam[+]
#>   latitude df  deg_free nparam[+]
```

Note that the `identifier` and `type` columns are not the same for both of the spline parameters.

When a recipe and model specification are combined using a workflow, both sets of parameters are shown:

```
wflow_param <-
  workflow() %>%
  add_recipe(ames_rec) %>%
  add_model(neural_net_spec) %>%
  extract_parameter_set_dials()
wflow_param
#> Collection of 4 parameters for tuning
#>
#>    identifier        type    object
#>  hidden_units hidden_units nparam[+]
#>     threshold    threshold nparam[+]
#>  longitude df     deg_free nparam[+]
#>   latitude df     deg_free nparam[+]
```

Neural networks are exquisitely capable of emulating nonlinear patterns. Adding spline terms to this type of model is unnecessary; we combined this model and recipe for illustration only.

Each tuning parameter argument has a corresponding function in the dials package. In the vast majority of the cases, the function has the same name as the parameter argument:

```
hidden_units()
#> # Hidden Units (quantitative)
#> Range: [1, 10]
threshold()
#> Threshold (quantitative)
#> Range: [0, 1]
```

The deg_free parameter is a counterexample; the notion of degrees of freedom comes up in a variety of different contexts. When used with splines, there is a specialized dials function called spline_degree() that is, by default, invoked for splines:

```
spline_degree()
#> Piecewise Polynomial Degree (quantitative)
#> Range: [1, 10]
```

The dials package also has a convenience function for extracting a particular parameter object:

```
# identify the parameter using the id value:
wflow_param %>% extract_parameter_dials("threshold")
#> Threshold (quantitative)
#> Range: [0, 0.1]
```

Inside the parameter set, the range of the parameters can also be updated in place:

```
extract_parameter_set_dials(ames_rec) %>%
  update(threshold = threshold(c(0.8, 1.0)))
#> Collection of 3 parameters for tuning
#>
#>    identifier      type    object
#>     threshold threshold nparam[+]
#>  longitude df   deg_free nparam[+]
#>   latitude df   deg_free nparam[+]
```

The *parameter sets* created by extract_parameter_set_dials() are consumed by the tidymodels tuning functions (when needed). If the defaults for the tuning parameter objects require modification, a modified parameter set is passed to the appropriate tuning function.

Some tuning parameters depend on the dimensions of the data. For example, the number of nearest neighbors must be between one and the number of rows in the data.

In some cases, it is easy to have reasonable defaults for the range of possible values. In other cases, the parameter range is critical and cannot be assumed. The primary tuning parameter for random forest models is the number of predictor columns that are randomly sampled for each split in the tree, usually denoted as mtry(). Without knowing the number of predictors, this parameter range cannot be preconfigured and requires finalization:

```
rf_spec <-
  rand_forest(mtry = tune()) %>%
  set_engine("ranger", regularization.factor = tune("regularization"))

rf_param <- extract_parameter_set_dials(rf_spec)
rf_param
#> Collection of 2 parameters for tuning
#>
#>       identifier                type    object
#>             mtry                mtry nparam[?]
#>   regularization regularization.factor nparam[+]
#>
#> Model parameters needing finalization:
#>    # Randomly Selected Predictors ('mtry')
#>
#> See `?dials::finalize` or `?dials::update.parameters` for more information.
```

Complete parameter objects have [+] in their summary; a value of [?] indicates that at least one end of the possible range is missing. There are two methods for handling this. The first is to use update() to add a range, based on what you know about the data dimensions:

```
rf_param %>%
  update(mtry = mtry(c(1, 70)))
#> Collection of 2 parameters for tuning
#>
#>       identifier                type    object
#>             mtry                mtry nparam[+]
#>   regularization regularization.factor nparam[+]
```

However, this approach might not work if a recipe is attached to a workflow that uses steps that either add or subtract columns. If those steps are not slated for tuning, the finalize() function can execute the recipe once to obtain the dimensions:

```
pca_rec <-
  recipe(Sale_Price ~ ., data = ames_train) %>%
  # Select the square-footage predictors and extract their PCA components:
  step_normalize(contains("SF")) %>%
  # Select the number of components needed to capture 95% of
  # the variance in the predictors.
  step_pca(contains("SF"), threshold = .95)
```

```
updated_param <-
  workflow() %>%
  add_model(rf_spec) %>%
  add_recipe(pca_rec) %>%
  extract_parameter_set_dials() %>%
  finalize(ames_train)
updated_param
#> Collection of 2 parameters for tuning
#>
#>        identifier                type    object
#>              mtry                mtry nparam[+]
#>   regularization regularization.factor nparam[+]
updated_param %>% extract_parameter_dials("mtry")
#> # Randomly Selected Predictors (quantitative)
#> Range: [1, 74]
```

When the recipe is prepared, the `finalize()` function learns to set the upper range of `mtry` to 74 predictors.

Additionally, the results of `extract_parameter_set_dials()` will include engine-specific parameters (if any). They are discovered in the same way as the main arguments and included in the parameter set. The dials package contains parameter functions for all potentially tunable engine-specific parameters:

```
rf_param
#> Collection of 2 parameters for tuning
#>
#>        identifier                type    object
#>              mtry                mtry nparam[?]
#>   regularization regularization.factor nparam[+]
#>
#> Model parameters needing finalization:
#>    # Randomly Selected Predictors ('mtry')
#>
#> See `?dials::finalize` or `?dials::update.parameters` for more information.
regularization_factor()
#> Gain Penalization (quantitative)
#> Range: [0, 1]
```

Finally, some tuning parameters are best associated with transformations. A good example of this is the penalty parameter associated with many regularized regression models. This parameter is nonnegative, and it is common to vary its values in log units. The primary dials parameter object indicates that a transformation is used by default:

```
penalty()
#> Amount of Regularization (quantitative)
#> Transformer: log-10 [1e-100, Inf]
#> Range (transformed scale): [-10, 0]
```

This is important to know, especially when altering the range. New range values must be in the transformed units:

```
# correct method to have penalty values between 0.1 and 1.0
penalty(c(-1, 0)) %>% value_sample(1000) %>% summary()
```

```
#>    Min. 1st Qu. Median    Mean 3rd Qu.    Max.
#>   0.101   0.181  0.327   0.400   0.589   0.999

# incorrect:
penalty(c(0.1, 1.0)) %>% value_sample(1000) %>% summary()
#>    Min. 1st Qu. Median    Mean 3rd Qu.    Max.
#>    1.26   2.21   3.68    4.26   5.89   10.00
```

The scale can be changed if desired with the `trans` argument. You can use natural units but the same range:

```
penalty(trans = NULL, range = 10^c(-10, 0))
#> Amount of Regularization (quantitative)
#> Range: [1e-10, 1]
```

Chapter Summary

This chapter introduced the process of tuning model hyperparameters that cannot be directly estimated from the data. Tuning such parameters can lead to overfitting, often by allowing a model to grow overly complex, so using resampled data sets together with appropriate metrics for evaluation is important. There are two general strategies for determining the right values, grid search and iterative search, which we will explore in depth in the next two chapters. In tidymodels, the `tune()` function is used to identify parameters for optimization, and functions from the dials package can extract and interact with tuning parameter objects.

## CHAPTER 13

Grid Search

In Chapter 12 we demonstrated how users can mark or tag arguments in preprocessing recipes and/or model specifications for optimization using the tune() function. Once we know what to optimize, it's time to address the question of how to optimize the parameters. This chapter describes *grid search* methods that specify the possible values of the parameters a priori. (Chapter 14 will continue the discussion by describing iterative search methods.)

Let's start by looking at two main approaches for assembling a grid.

Regular and Nonregular Grids

There are two main types of grids. A regular grid combines each parameter (with its corresponding set of possible values) factorially, i.e., by using all combinations of the sets. Alternatively, a nonregular grid is one where the parameter combinations are not formed from a small set of points.

Before we look at each type in more detail, let's consider an example model: the multilayer perceptron model (a.k.a. single-layer artificial neural network). The parameters marked for tuning are:

- The number of hidden units
- The number of fitting epochs/iterations in model training
- The amount of weight decay penalization

Using parsnip, the specification for a classification model fit using the nnet package is:

```
library(tidymodels)
tidymodels_prefer()

mlp_spec <-
  mlp(hidden_units = tune(), penalty = tune(), epochs = tune()) %>%
  set_engine("nnet", trace = 0) %>%
  set_mode("classification")
```

The argument `trace = 0` prevents extra logging of the training process. As shown in Chapter 12, the `extract_parameter_set_dials()` function can extract the set of arguments with unknown values and set their dials objects:

```
mlp_param <- extract_parameter_set_dials(mlp_spec)
mlp_param %>% extract_parameter_dials("hidden_units")
#> # Hidden Units (quantitative)
#> Range: [1, 10]
mlp_param %>% extract_parameter_dials("penalty")
#> Amount of Regularization (quantitative)
#> Transformer: log-10 [1e-100, Inf]
#> Range (transformed scale): [-10, 0]
mlp_param %>% extract_parameter_dials("epochs")
#> # Epochs (quantitative)
#> Range: [10, 1000]
```

This output indicates that the parameter objects are complete and prints their default ranges. These values will be used to demonstrate how to create different types of parameter grids.

 Historically, the number of epochs was determined by early stopping; a separate validation set determined the length of training based on the error rate, since repredicting the training set led to overfitting. In our case, the use of a weight decay penalty should prohibit overfitting, and there is little harm in tuning the penalty and the number of epochs.

Regular Grids

Regular grids are combinations of separate sets of parameter values. First, the user creates a distinct set of values for each parameter. The number of possible values need not be the same for each parameter. The tidyr function `crossing()` is one way to create a regular grid:

```
crossing(
  hidden_units = 1:3,
  penalty = c(0.0, 0.1),
  epochs = c(100, 200)
)
#> # A tibble: 12 × 3
#>   hidden_units penalty epochs
#>          <int>   <dbl>  <dbl>
```

```
#> 1                  1      0      100
#> 2                  1      0      200
#> 3                  1      0.1    100
#> 4                  1      0.1    200
#> 5                  2      0      100
#> 6                  2      0      200
#> # … with 6 more rows
```

The parameter object knows the ranges of the parameters. The dials package contains a set of grid_*() functions that take the parameter object as input to produce different types of grids. For example:

```
grid_regular(mlp_param, levels = 2)
#> # A tibble: 8 × 3
#>   hidden_units        penalty epochs
#>          <int>          <dbl>  <int>
#> 1              1 0.0000000001     10
#> 2             10 0.0000000001     10
#> 3              1 1                10
#> 4             10 1                10
#> 5              1 0.0000000001   1000
#> 6             10 0.0000000001   1000
#> # … with 2 more rows
```

The levels argument is the number of levels per parameter to create. It can also take a named vector of values:

```
mlp_param %>%
  grid_regular(levels = c(hidden_units = 3, penalty = 2, epochs = 2))
#> # A tibble: 12 × 3
#>   hidden_units        penalty epochs
#>          <int>          <dbl>  <int>
#> 1              1 0.0000000001     10
#> 2              5 0.0000000001     10
#> 3             10 0.0000000001     10
#> 4              1 1                10
#> 5              5 1                10
#> 6             10 1                10
#> # … with 6 more rows
```

There are techniques for creating regular grids that do not use all possible values of each parameter set. These *fractional factorial designs* (Box, Hunter, and Hunter 2005) could also be used. To learn more, consult the CRAN Task View for experimental design (*https://oreil.ly/PvLCj*).

> Regular grids can be computationally expensive to use, especially when there are a medium-to-large number of tuning parameters. This is true for many models but not all. As discussed further in this chapter, there are many models whose tuning time *decreases* with a regular grid!

One advantage to using a regular grid is that the relationships and patterns between the tuning parameters and the model metrics are easily understood. The factorial nature of these designs allows for examination of each parameter separately with little confounding between parameters.

Nonregular Grids

There are several options for creating nonregular grids. The first is to use random sampling across the range of parameters. The `grid_random()` function generates independent uniform random numbers across the parameter ranges. If the parameter object has an associated transformation (such as we have for `penalty`), the random numbers are generated on the transformed scale. Let's create a random grid for the parameters from our example neural network:

```
set.seed(1301)
mlp_param %>%
  grid_random(size = 1000) %>% # 'size' is the number of combinations
  summary()
#>    hidden_units       penalty           epochs
#>   Min.   : 1.00   Min.   :0.0000   Min.   : 10
#>   1st Qu.: 3.00   1st Qu.:0.0000   1st Qu.:266
#>   Median : 5.00   Median :0.0000   Median :497
#>   Mean   : 5.38   Mean   :0.0437   Mean   :510
#>   3rd Qu.: 8.00   3rd Qu.:0.0027   3rd Qu.:761
#>   Max.   :10.00   Max.   :0.9814   Max.   :999
```

For `penalty`, the random numbers are uniform on the log (base-10) scale, but the values in the grid are in the natural units.

The issue with random grids is that, with small-to-medium grids, random values can result in overlapping parameter combinations. Also, the random grid needs to cover the whole parameter space, but the likelihood of good coverage increases with the number of grid values. Even for a sample of 15 candidate points, Figure 13-1 shows some overlap between points for our example multilayer perceptron:

```
library(ggforce)
set.seed(1302)
mlp_param %>%
  # The 'original = FALSE' option keeps penalty in log10 units
  grid_random(size = 20, original = FALSE) %>%
  ggplot(aes(x = .panel_x, y = .panel_y)) +
  geom_point() +
  geom_blank() +
  facet_matrix(vars(hidden_units, penalty, epochs), layer.diag = 2) +
  labs(title = "Random design with 20 candidates")
```

A much better approach is to use a set of experimental designs called *space-filling designs*. While different design methods have slightly different goals, they generally find a configuration of points that cover the parameter space with the smallest chance of overlapping or redundant values. Examples of such designs are Latin hypercubes (McKay, Beckman, and Conover 1979), maximum entropy designs (Shewry and Wynn 1987), maximum projection designs (Joseph, Gul, and Ba 2015), and others. See Santner et al. (2003) for an overview.

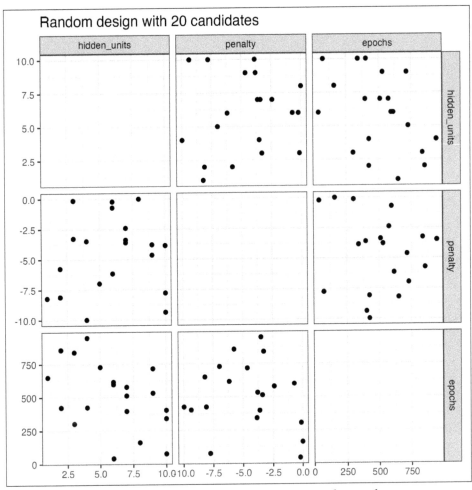

Figure 13-1. Three tuning parameters with 20 points generated at random.

The dials package contains functions for Latin hypercube and maximum entropy designs. As with grid_random(), the primary inputs are the number of parameter combinations and a parameter object. Let's compare a random design with a Latin hypercube design for 15 candidate parameter values in Figure 13-2:

```
set.seed(1303)
mlp_param %>%
  grid_latin_hypercube(size = 20, original = FALSE) %>%
  ggplot(aes(x = .panel_x, y = .panel_y)) +
  geom_point() +
  geom_blank() +
  facet_matrix(vars(hidden_units, penalty, epochs), layer.diag = 2) +
  labs(title = "Latin Hypercube design with 20 candidates")
```

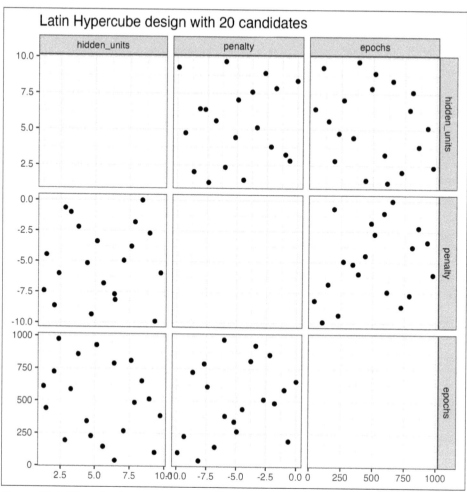

Figure 13-2. Three tuning parameters with 20 points generated using a space-filling design.

While not perfect, this Latin hypercube design spaces the points farther away from one another and allows a better exploration of the hyperparameter space.

Space-filling designs can be very effective at representing the parameter space. The default design used by the tune package is the maximum entropy design. These tend to produce grids that cover the candidate space well and drastically increase the chances of finding good results.

Evaluating the Grid

To choose the best tuning parameter combination, each candidate set is assessed using data that were not used to train that model. Resampling methods or a single validation set work well for this purpose. The process (and syntax) closely resembles the approach in Chapter 10 that used the `fit_resamples()` function from the tune package.

After resampling, the user selects the most appropriate candidate parameter set. It might make sense to choose the empirically best parameter combination or bias the choice toward other aspects of the model fit, such as simplicity.

We use a classification data set to demonstrate model tuning in this and the next chapter. The data come from Hill et al. (2007), who developed an automated microscopy laboratory tool for cancer research. The data consist of 56 imaging measurements on 2,019 human breast cancer cells. These predictors represent shape and intensity characteristics of different parts of the cells (e.g., the nucleus, the cell boundary, etc.). There is a high degree of correlation between the predictors. For example, there are several different predictors that measure the size and shape of the nucleus and cell boundary. Also, individually, many predictors have skewed distributions.

Each cell belongs to one of two classes. Since this is part of an automated lab test, the focus was on prediction capability rather than inference.

The data are included in the modeldata package. Let's remove one column not needed for analysis (`case`):

```
library(tidymodels)
data(cells)
cells <- cells %>% select(-case)
```

Given the dimensions of the data, we can compute performance metrics using 10-fold cross-validation:

```
set.seed(1304)
cell_folds <- vfold_cv(cells)
```

Because of the high degree of correlation between predictors, it makes sense to use PCA feature extraction to decorrelate the predictors. The following recipe contains steps to transform the predictors to increase symmetry, normalize them to be on the same scale, and then conduct feature extraction. The number of PCA components to retain is also tuned, along with the model parameters.

 While the resulting PCA components are technically on the same scale, the lower-rank components tend to have a wider range than the higher-rank components. For this reason, we normalize again to coerce the predictors to have the same mean and variance.

Many of the predictors have skewed distributions. Since PCA is variance based, extreme values can have a detrimental effect on these calculations. To counter this, let's add a recipe step estimating a Yeo–Johnson transformation for each predictor (Yeo and Johnson 2000). While originally intended as a transformation of the outcome, it can also be used to estimate transformations that encourage more symmetric distributions. This step, `step_YeoJohnson()`, occurs in the recipe just prior to the initial normalization via `step_normalize()`. Then, let's combine this feature engineering recipe with our neural network model specification `mlp_spec`:

```
mlp_rec <-
  recipe(class ~ ., data = cells) %>%
  step_YeoJohnson(all_numeric_predictors()) %>%
  step_normalize(all_numeric_predictors()) %>%
  step_pca(all_numeric_predictors(), num_comp = tune()) %>%
  step_normalize(all_numeric_predictors())

mlp_wflow <-
  workflow() %>%
  add_model(mlp_spec) %>%
  add_recipe(mlp_rec)
```

Let's create a parameter object `mlp_param` to adjust a few of the default ranges. We can change the number of epochs to have a smaller range (50 to 200 epochs). Also, the default range for `num_comp()` defaults to a very narrow range (one to four components); we can increase the range to 40 components and set the minimum value to zero:

```
mlp_param <-
  mlp_wflow %>%
  extract_parameter_set_dials() %>%
  update(
    epochs = epochs(c(50, 200)),
    num_comp = num_comp(c(0, 40))
  )
```

In `step_pca()`, using zero PCA components is a shortcut to skip the feature extraction. In this way, the original predictors can be directly compared to the results that include PCA components.

The `tune_grid()` function is the primary function for conducting grid search. Its functionality is very similar to `fit_resamples()`, although it has additional arguments related to the grid:

grid

An integer or data frame. When an integer is used, the function creates a space-filling design with `grid` number of candidate parameter combinations. If specific parameter combinations exist, the `grid` parameter is used to pass them to the function.

`param_info`

    An optional argument for defining the parameter ranges. The argument is most useful when `grid` is an integer.

Otherwise, the interface to `tune_grid()` is the same as `fit_resamples()`. The first argument is either a model specification or workflow. When a model is given, the second argument can be either a recipe or a formula. The other required argument is an rsample resampling object (such as `cell_folds`). The following call also passes a metric set so that the area under the ROC curve is measured during resampling.

To start, let's evaluate a regular grid with three levels across the resamples:

```
roc_res <- metric_set(roc_auc)
set.seed(1305)
mlp_reg_tune <-
  mlp_wflow %>%
  tune_grid(
    cell_folds,
    grid = mlp_param %>% grid_regular(levels = 3),
    metrics = roc_res
  )
mlp_reg_tune
#> # Tuning results
#> # 10-fold cross-validation
#> # A tibble: 10 × 4
#>    splits             id     .metrics        .notes
#>    <list>             <chr>  <list>          <list>
#> 1 <split [1817/202]> Fold01 <tibble [81 × 8]> <tibble [0 × 3]>
#> 2 <split [1817/202]> Fold02 <tibble [81 × 8]> <tibble [0 × 3]>
#> 3 <split [1817/202]> Fold03 <tibble [81 × 8]> <tibble [0 × 3]>
#> 4 <split [1817/202]> Fold04 <tibble [81 × 8]> <tibble [0 × 3]>
#> 5 <split [1817/202]> Fold05 <tibble [81 × 8]> <tibble [0 × 3]>
#> 6 <split [1817/202]> Fold06 <tibble [81 × 8]> <tibble [0 × 3]>
#> # … with 4 more rows
```

There are high-level convenience functions we can use to understand the results. First, the `autoplot()` method for regular grids shows the performance profiles across tuning parameters in Figure 13-3:

```
autoplot(mlp_reg_tune) +
  scale_color_viridis_d(direction = -1) +
  theme(legend.position = "top")
```

For these data, the amount of penalization has the largest impact on the area under the ROC curve. The number of epochs doesn't appear to have a pronounced effect on performance. The change in the number of hidden units appears to matter most when the amount of regularization is low (and harms performance). There are several parameter configurations that have roughly equivalent performance, as seen using the function `show_best()`:

```
show_best(mlp_reg_tune) %>% select(-.estimator)
#> # A tibble: 5 × 9
#>   hidden_units penalty epochs num_comp .metric mean    n std_err .config
#>          <int>   <dbl>  <int>    <int> <chr>   <dbl> <int>  <dbl> <chr>
```

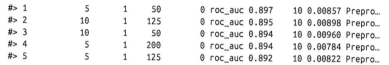

```
#> 1         5     1     50       0 roc_auc 0.897    10 0.00857 Prepro...
#> 2        10     1    125       0 roc_auc 0.895    10 0.00898 Prepro...
#> 3        10     1     50       0 roc_auc 0.894    10 0.00960 Prepro...
#> 4         5     1    200       0 roc_auc 0.894    10 0.00784 Prepro...
#> 5         5     1    125       0 roc_auc 0.892    10 0.00822 Prepro...
```

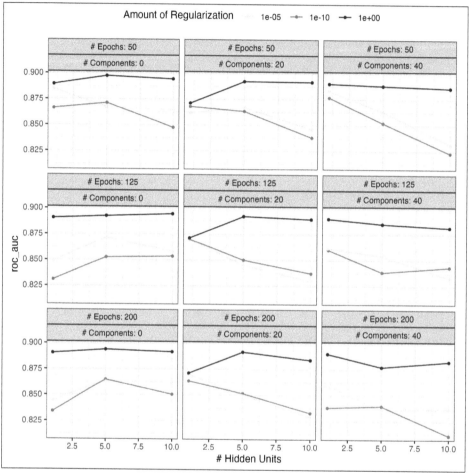

Figure 13-3. The regular grid results.

Based on these results, it would make sense to conduct another run of grid search with larger values of the weight decay penalty.

To use a space-filling design, either the grid argument can be given an integer or one of the grid_*() functions can produce a data frame. To evaluate the same range using a maximum entropy design with 20 candidate values:

```
set.seed(1306)
mlp_sfd_tune <-
  mlp_wflow %>%
```

```
  tune_grid(
    cell_folds,
    grid = 20,
    # Pass in the parameter object to use the appropriate range:
    param_info = mlp_param,
    metrics = roc_res
  )
mlp_sfd_tune
#> # Tuning results
#> # 10-fold cross-validation
#> # A tibble: 10 × 4
#>   splits              id     .metrics           .notes
#>   <list>              <chr>  <list>             <list>
#> 1 <split [1817/202]>  Fold01 <tibble [20 × 8]>  <tibble [0 × 3]>
#> 2 <split [1817/202]>  Fold02 <tibble [20 × 8]>  <tibble [0 × 3]>
#> 3 <split [1817/202]>  Fold03 <tibble [20 × 8]>  <tibble [0 × 3]>
#> 4 <split [1817/202]>  Fold04 <tibble [20 × 8]>  <tibble [0 × 3]>
#> 5 <split [1817/202]>  Fold05 <tibble [20 × 8]>  <tibble [0 × 3]>
#> 6 <split [1817/202]>  Fold06 <tibble [20 × 8]>  <tibble [0 × 3]>
#> # … with 4 more rows
```

The `autoplot()` method will also work with these designs, although the format of the results will be different. Figure 13-4 was produced using `autoplot(mlp_sfd_tune)`.

This marginal effects plot (Figure 13-4) shows the relationship of each parameter with the performance metric.

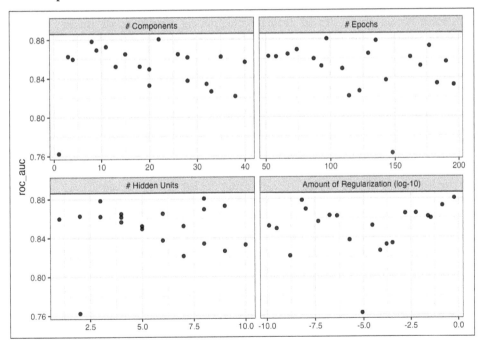

Figure 13-4. The `autoplot()` method results when used with a space-filling design.

Take care when examining this plot; since a regular grid is not used, the values of the other tuning parameters can affect each panel.

The penalty parameter appears to result in better performance with smaller amounts of weight decay. This is the opposite of the results from the regular grid. Since each point in each panel is shared with the other three tuning parameters, the trends in one panel can be affected by the others. Using a regular grid, each point in each panel is equally averaged over the other parameters. For this reason, the effect of each parameter is better isolated with regular grids.

As with the regular grid, show_best() can report on the numerically best results:

```
show_best(mlp_sfd_tune) %>% select(-.estimator)
#> # A tibble: 5 × 9
#>   hidden_units       penalty epochs num_comp .metric mean      n std_err .config
#>          <int>         <dbl>  <int>    <int> <chr>   <dbl> <int>   <dbl> <chr>
#> 1            8 0.594             97       22 roc_auc 0.880    10 0.00998 Preprocess…
#> 2            3 0.00000000649    135        8 roc_auc 0.878    10 0.00956 Preprocess…
#> 3            9 0.141            177       11 roc_auc 0.873    10 0.0104  Preprocess…
#> 4            8 0.0000000103      74        9 roc_auc 0.869    10 0.00761 Preprocess…
#> 5            6 0.00581          129       15 roc_auc 0.865    10 0.00658 Preprocess…
```

Generally, it is a good idea to evaluate the models over multiple metrics so that different aspects of the model fit are taken into account. Also, it often makes sense to choose a slightly suboptimal parameter combination that is associated with a simpler model. For this model, simplicity corresponds to larger penalty values and/or fewer hidden units.

As with the results from fit_resamples(), there is usually no value in retaining the intermediary model fits across the resamples and tuning parameters. However, as before, the extract option to control_grid() allows the retention of the fitted models and/or recipes. Also, setting the save_pred option to TRUE retains the assessment set predictions and these can be accessed using collect_predictions().

Finalizing the Model

If one of the sets of possible model parameters found via show_best() were an attractive final option for these data, we might wish to evaluate how well it does on the test set. However, the results of tune_grid() only provide the substrate to choose appropriate tuning parameters. The function *does not fit* a final model.

To fit a final model, a final set of parameter values must be determined. There are two methods to do so:

- manually pick values that appear appropriate or

- use a select_*() function.

For example, select_best() will choose the parameters with the numerically best results. Let's go back to our regular grid results and see which one is best:

```
select_best(mlp_reg_tune, metric = "roc_auc")
#> # A tibble: 1 × 5
#>   hidden_units penalty epochs num_comp .config
#>          <int>   <dbl>  <int>    <int> <chr>
#> 1            5       1     50        0 Preprocessor1_Model08
```

Looking back at Figure 13-3, we can see that a model with a single hidden unit trained for 125 epochs on the original predictors with a large amount of penalization performs competitively with this option and is simpler. This is basically penalized logistic regression! To manually specify these parameters, we can create a tibble with these values and then use a *finalization* function to splice the values back into the workflow:

```
logistic_param <-
  tibble(
    num_comp = 0,
    epochs = 125,
    hidden_units = 1,
    penalty = 1
  )

final_mlp_wflow <-
  mlp_wflow %>%
  finalize_workflow(logistic_param)
final_mlp_wflow
#> ══ Workflow ═══════════════════════════════════
#> Preprocessor: Recipe
#> Model: mlp()
#>
#> ── Preprocessor ───────────────────────────────
#> 4 Recipe Steps
#>
#> • step_YeoJohnson()
#> • step_normalize()
#> • step_pca()
#> • step_normalize()
#>
#> ── Model ──────────────────────────────────────
#> Single Layer Neural Network Specification (classification)
#>
#> Main Arguments:
#>   hidden_units = 1
#>   penalty = 1
#>   epochs = 125
#>
#> Engine-Specific Arguments:
#>   trace = 0
#>
#> Computational engine: nnet
```

No more values of `tune()` are included in this finalized workflow. Now the model can be fit to the entire training set:

```
final_mlp_fit <-
  final_mlp_wflow %>%
  fit(cells)
```

This object can now be used to make future predictions on new data.

If you did not use a workflow, finalization of a model and/or recipe is done using `finalize_model()` and `finalize_recipe()`.

Tools for Creating Tuning Specifications

The usemodels package can take a data frame and model formula, then write out R code for tuning the model. The code also creates an appropriate recipe whose steps depend on the requested model as well as the predictor data. For example, for the Ames housing data, `xgboost` modeling code could be created with:

```
library(usemodels)

use_xgboost(Sale_Price ~ Neighborhood + Gr_Liv_Area + Year_Built + Bldg_Type +
              Latitude + Longitude,
            data = ames_train,
            # Add comments explaining some of the code:
            verbose = TRUE)
```

The resulting code is as follows:

```
xgboost_recipe <-
  recipe(formula = Sale_Price ~ Neighborhood + Gr_Liv_Area + Year_Built + Bldg_Type +
    Latitude + Longitude, data = ames_train) %>%
  step_novel(all_nominal_predictors()) %>%
  ## This model requires the predictors to be numeric. The most common
  ## method to convert qualitative predictors to numeric is to create
  ## binary indicator variables (aka dummy variables) from these
  ## predictors. However, for this model, binary indicator variables can be
  ## made for each of the levels of the factors (known as 'one-hot
  ## encoding').
  step_dummy(all_nominal_predictors(), one_hot = TRUE) %>%
  step_zv(all_predictors())

xgboost_spec <-
  boost_tree(trees = tune(), min_n = tune(), tree_depth = tune(), learn_rate = tune(),
    loss_reduction = tune(), sample_size = tune()) %>%
  set_mode("regression") %>%
  set_engine("xgboost")

xgboost_workflow <-
  workflow() %>%
  add_recipe(xgboost_recipe) %>%
  add_model(xgboost_spec)

set.seed(69305)
xgboost_tune <-
  tune_grid(xgboost_workflow,
```

```
    resamples = stop("add your rsample object"),
    grid = stop("add number of candidate points"))
```

Based on what usemodels understands about the data, this code is the minimal preprocessing required. For other models, operations like `step_normalize()` are added to fulfill the basic needs of the model. Notice that it is our responsibility, as the modeling practitioner, to choose what `resamples` to use for tuning, as well as what kind of `grid`.

 The usemodels package can also be used to create model fitting code with no tuning by setting the argument `tune = FALSE`.

Tools for Efficient Grid Search

It is possible to make grid search more computationally efficient by applying a few different tricks and optimizations. This section describes several techniques.

Submodel Optimization

There are types of models where, from a single model fit, multiple tuning parameters can be evaluated without refitting.

For example, partial least squares (PLS) is a supervised version of principal component analysis (Geladi and Kowalski 1986). It creates components that maximize the variation in the predictors (like PCA) but simultaneously tries to maximize the correlation between these predictors and the outcome. We'll explore PLS more in Chapter 16. One tuning parameter is the number of PLS components to retain. Suppose that a data set with 100 predictors is fit using PLS. The number of possible components to retain can range from 1 to 50. However, in many implementations, a single model fit can compute predicted values across many values of `num_comp`. As a result, a PLS model created with 100 components can also make predictions for any `num_comp <= 100`. This saves time since, instead of creating redundant model fits, a single fit can be used to evaluate many submodels.

While not all models can exploit this feature, many broadly used ones do.

- Boosting models can typically make predictions across multiple values for the number of boosting iterations.

- Regularization methods, such as the glmnet model, can make simultaneous predictions across the amount of regularization used to fit the model.

- Multivariate adaptive regression splines (MARS) adds a set of nonlinear features to linear regression models (Friedman 1991). The number of terms to retain is a tuning parameter, and it is computationally fast to make predictions across many values of this parameter from a single model fit.

The tune package automatically applies this type of optimization whenever an applicable model is tuned.

For example, if a boosted C5.0 classification model (Kuhn and Johnson 2013) were fit to the cell data, we could tune the number of boosting iterations (trees). With all other parameters set at their default values, we could evaluate iterations from 1 to 100 on the same resamples as used previously:

```
c5_spec <-
  boost_tree(trees = tune()) %>%
  set_engine("C5.0") %>%
  set_mode("classification")

set.seed(1307)
c5_spec %>%
  tune_grid(
    class ~ .,
    resamples = cell_folds,
    grid = data.frame(trees = 1:100),
    metrics = roc_res
  )
```

Without the submodel optimization, the call to tune_grid() used 62.2 minutes to resample 100 submodels. With the optimization, the same call took 100 *seconds* (a 37-fold speedup). The reduced time is the difference in tune_grid() fitting 1,000 models versus 10 models.

Even though we fit the model with and without the submodel prediction trick, this optimization is automatically applied by parsnip.

Parallel Processing

As previously mentioned in Chapter 10, parallel processing is an effective method for decreasing execution time when resampling models. This advantage conveys to model tuning via grid search, although there are additional considerations.

Let's consider two different parallel processing schemes.

When tuning models via grid search, there are two distinct loops: one over resamples and another over the unique tuning parameter combinations. In pseudocode, this process would look like:

```
for (rs in resamples) {
  # Create analysis and assessment sets
  # Preprocess data (e.g. formula or recipe)
```

```
for (mod in configurations) {
  # Fit model {mod} to the {rs} analysis set
  # Predict the {rs} assessment set
  }
}
```

By default, the tune package parallelizes only over resamples (the outer loop), as opposed to both the outer and inner loops.

This is the optimal scenario when the preprocessing method is expensive. However, there are two potential downsides to this approach:

- It limits the achievable speed-ups when the preprocessing is not expensive.
- The number of parallel workers is limited by the number of resamples. For example, with 10-fold cross-validation you can use only 10 parallel workers even when the computer has more than 10 cores.

To illustrate how the parallel processing works, we'll use a case where there are seven model tuning parameter values, with 5-fold cross-validation. Figure 13-5 shows how the tasks are allocated to the worker processes.

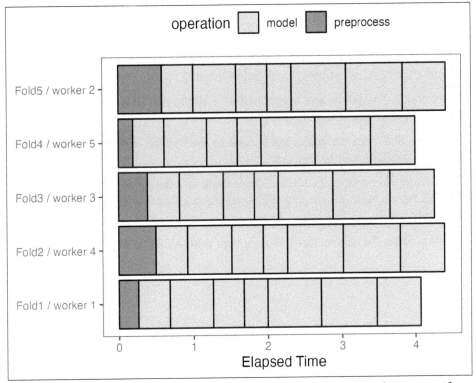

Figure 13-5. Worker processes when parallel processing matches resamples to a specific worker process.

Note that each fold is assigned to its own worker process and, since only model parameters are being tuned, the preprocessing is conducted once per fold/worker. If fewer than five worker processes were used, some workers would receive multiple folds.

In the control functions for the `tune_*()` functions, the argument `parallel_over` controls how the process is executed. To use the previous parallelization strategy, the argument is `parallel_over = "resamples"`.

Instead of parallel processing the resamples, an alternate scheme combines the loops over resamples and models into a single loop. In pseudocode, this process would look like:

```
all_tasks <- crossing(resamples, configurations)

for (iter in all_tasks) {
  # Create analysis and assessment sets for {iter}
  # Preprocess data (e.g. formula or recipe)
  # Fit model {iter} to the {iter} analysis set
  # Predict the {iter} assessment set
}
```

In this case, parallelization now occurs over the single loop. For example, if we use 5-fold cross-validation with M tuning parameter values, the loop is executed over $5 \times M$ iterations. This increases the number of potential workers that can be used. However, the work related to data preprocessing is repeated multiple times. If those steps are expensive, this approach will be inefficient.

In tidymodels, validation sets are treated as a single resample. In these cases, this parallelization scheme would be best.

Figure 13-6 illustrates the delegation of tasks to the workers in this scheme; the same example is used but with 10 workers.

Here, each worker process handles multiple folds, and the preprocessing is needlessly repeated. For example, for the first fold, the preprocessing was computed seven times instead of once.

For this scheme, the control function argument is `parallel_over = "everything"`.

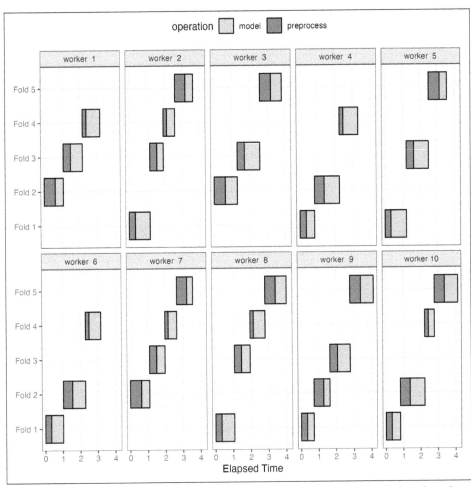

Figure 13-6. Worker processes when preprocessing and modeling tasks are distributed to many workers.

Benchmarking Boosted Trees

To compare different possible parallelization schemes, we tuned a boosted tree with the xgboost engine using a data set of 4,000 samples, with 5-fold cross-validation and 10 candidate models. These data required some baseline preprocessing that did not require any estimation. The preprocessing was handled three different ways:

1. Preprocess the data prior to modeling using a dplyr pipeline (labeled as "none" in the later plots).

2. Conduct the same preprocessing via a recipe (shown as "light" preprocessing).

3. With a recipe, add an additional step that has a high computational cost (labeled as "expensive").

The first and second preprocessing options are designed for comparison, to measure the computational cost of the recipe in the second option. The third option measures the cost of performing redundant computations with `parallel_over` = `"everything"`.

We evaluated this process using variable numbers of worker processes and using the two `parallel_over` options, on a computer with 10 physical cores and 20 virtual cores (via hyperthreading).

First, let's consider the raw execution times in Figure 13-7.

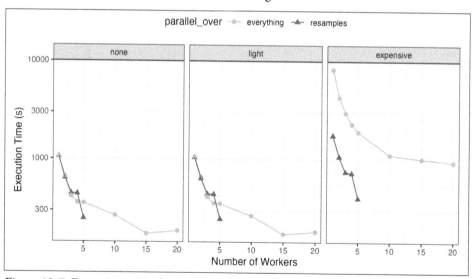

Figure 13-7. Execution times for model tuning versus the number of workers using different delegation schemes. The diagonal black line indicates a linear speedup where the addition of a new worker process has maximal effect.

Since there were only five resamples, the number of cores used when `parallel_over` = "resamples" is limited to five.

Comparing the curves in the first two panels for "none" and "light":

- There is little difference in the execution times between the panels. This indicates, for these data, there is no real computational penalty for doing the preprocessing steps in a recipe.
- There is some benefit for using `parallel_over` = "everything" with many cores. However, as shown in the figure, the majority of the benefit of parallel processing occurs in the first five workers.

With the expensive preprocessing step, there is a considerable difference in execution times. Using `parallel_over` = "everything" is problematic since, even using all cores, it never achieves the execution time that `parallel_over` = "resamples" attains with just five cores. This is because the costly preprocessing step is unnecessarily repeated in the computational scheme.

We can also view these data in terms of speed-ups in Figure 13-8.

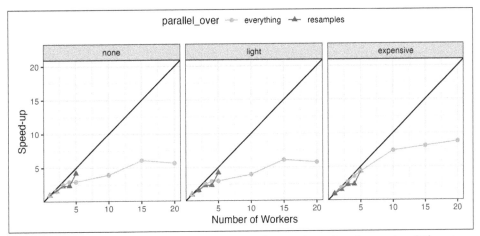

Figure 13-8. Speed-ups for model tuning versus the number of workers using different delegation schemes.

The best speed-ups, for these data, occur when `parallel_over` = "resamples" and when the computations are expensive. However, in the latter case, remember that the previous analysis indicates that the overall model fits are slower.

What is the benefit of using the submodel optimization method in conjunction with parallel processing? The C5.0 classification model shown previously in this chapter was also run in parallel with 10 workers. The parallel computations took 13.3 seconds for a 7.5-fold speed-up (both runs used the submodel optimization trick). Between the submodel optimization trick and parallel processing, there was a total 282-fold speed-up over the most basic grid search code.

Overall, note that the increased computational savings will vary from model to model and are also affected by the size of the grid, the number of resamples, etc. A very computationally efficient model may not benefit as much from parallel processing.

Access to Global Variables

When using tidymodels, it is possible to use values in your local environment (usually the global environment) in model objects.

What do we mean by "environment" here? Think of an environment in R as a place to store variables that you can work with. See the "Environments" chapter of Wickham (2019) to learn more.

If we define a variable to use as a model parameter and then pass it to a function like `linear_reg()`, the variable is typically defined in the global environment:

```
coef_penalty <- 0.1
spec <- linear_reg(penalty = coef_penalty) %>% set_engine("glmnet")
spec
#> Linear Regression Model Specification (regression)
#>
#> Main Arguments:
#>   penalty = coef_penalty
#>
#> Computational engine: glmnet
```

Models created with the parsnip package save arguments like these as *quosures*; these are objects that track both the name of the object as well as the environment where it lives:

```
spec$args$penalty
#> <quosure>
#> expr: ^coef_penalty
#> env:  global
```

Notice that we have `env: global` because this variable was created in the global environment. The model specification defined by `spec` works correctly when run in a user's regular session because that session is also using the global environment; R can easily find the object `coef_penalty`.

When such a model is evaluated with parallel workers, it may fail. Depending on the particular technology that is used for parallel processing, the workers may not have access to the global environment.

When writing code that will be run in parallel, it is a good idea to insert the actual data into the objects rather than the reference to the object. The rlang and dplyr packages can be very helpful for this. For example, the !! operator can splice a single value into an object:

```
spec <- linear_reg(penalty = !!coef_penalty) %>% set_engine("glmnet")
spec$args$penalty
#> <quosure>
#> expr: ^0.1
#> env:  empty
```

Now the output is ^0.1, indicating that the value is there instead of the reference to the object. When you have multiple external values to insert into an object, the !!! operator can help:

```
mcmc_args <- list(chains = 3, iter = 1000, cores = 3)

linear_reg() %>% set_engine("stan", !!!mcmc_args)
#> Linear Regression Model Specification (regression)
#>
#> Engine-Specific Arguments:
#>   chains = 3
#>   iter = 1000
#>   cores = 3
#>
#> Computational engine: stan
```

Recipe selectors are another place where you might want access to global variables. Suppose you have a recipe step that should use all of the predictors in the cell data that were measured using the second optical channel. We can create a vector of these column names:

```
library(stringr)
ch_2_vars <- str_subset(names(cells), "ch_2")
ch_2_vars
#> [1] "avg_inten_ch_2"   "total_inten_ch_2"
```

We could hard-code these into a recipe step, but it would be better to reference them programmatically in case the data change. Two ways to do this are:

```
# Still uses a reference to global data (~_~;)
recipe(class ~ ., data = cells) %>%
  step_spatialsign(all_of(ch_2_vars))
#> Recipe
#>
#> Inputs:
#>
#>         role #variables
#>      outcome          1
```

```
#>   predictor          56
#>
#> Operations:
#>
#> Spatial sign on  all_of(ch_2_vars)

# Inserts the values into the step  \(•_•)/
recipe(class ~ ., data = cells) %>%
  step_spatialsign(!!!ch_2_vars)
#> Recipe
#>
#> Inputs:
#>
#>         role #variables
#>      outcome          1
#>    predictor         56
#>
#> Operations:
#>
#> Spatial sign on  "avg_inten_ch_2", "total_inten_ch_2"
```

The latter is better for parallel processing because all of the needed information is embedded in the recipe object.

Racing Methods

One issue with grid search is that all models need to be fit across all resamples before any tuning parameters can be evaluated. It would be helpful if instead, at some point during tuning, an interim analysis could be conducted to eliminate any truly awful parameter candidates. This would be akin to *futility analysis* in clinical trials. If a new drug is performing excessively poorly (or well), it is potentially unethical to wait until the trial finishes to make a decision.

In machine learning, techniques called *racing methods* provide a similar function (Maron and Moore 1993). Here, the tuning process evaluates all models on an initial subset of resamples. Based on their current performance metrics, some parameter sets are not considered in subsequent resamples.

As an example, in the multilayer perceptron tuning process with a regular grid explored in this chapter, what would the results look like after only the first three folds? Using techniques similar to those shown in Chapter 11, we can fit a model where the outcome is the resampled area under the ROC curve and the predictor is an indicator for the parameter combination. The model takes the resample-to-resample effect into account and produces point and interval estimates for each parameter setting. The results of the model are one-sided 95% confidence intervals that measure the loss of the ROC value relative to the currently best performing parameters.

Figure 13-9 shows the results at several iterations in the process. The points shown in the panel with the first iteration show single ROC AUC values. As iterations progress, the points are averages of the resampled ROC statistics.

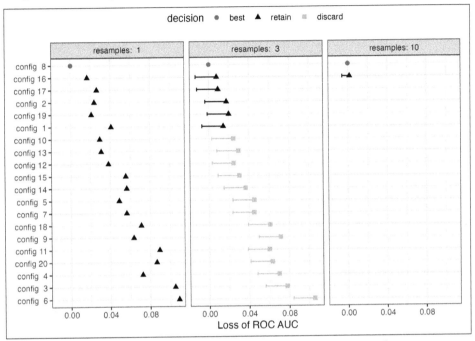

Figure 13-9. The racing process for 20 tuning parameters and 10 resamples.

On the third iteration, the leading model configuration has changed and the algorithm computes one-sided confidence intervals. Any parameter set whose confidence interval includes 0 would lack evidence that its performance is not statistically different from the best results. We retain 14 settings; these are resampled more. The remaining six submodels are no longer considered.

The process continues to resample configurations that remain, and the statistical analysis repeats with the current results. More submodels may be removed from consideration. Prior to the final resample, almost all submodels are eliminated and, at the last iteration, only two remain.[1]

---

1 See Max Kuhn (2014) for more details on the computational aspects of this approach.

Racing methods can be more efficient than basic grid search as long as the interim analysis is fast and some parameter settings have poor performance. It also is most helpful when the model does *not* have the ability to exploit submodel predictions.

The finetune package contains functions for racing. The `tune_race_anova()` function conducts an ANOVA model to test for statistical significance of the different model configurations. The syntax to reproduce the filtering shown previously is:

```
library(finetune)

set.seed(1308)
mlp_sfd_race <-
  mlp_wflow %>%
  tune_race_anova(
    cell_folds,
    grid = 20,
    param_info = mlp_param,
    metrics = roc_res,
    control = control_race(verbose_elim = TRUE)
  )
```

The arguments mirror those of `tune_grid()`. The function `control_race()` has options for the elimination procedure.

As shown in Figure 13-9, there were two tuning parameter combinations under consideration once the full set of resamples were evaluated. `show_best()` returns the best models (ranked by performance) but returns only the configurations that were never eliminated:

```
show_best(mlp_sfd_race, n = 10)
#> # A tibble: 2 × 10
#>   hidden_units penalty epochs num_comp .metric .estimator   mean     n std_err
#>          <int>   <dbl>  <int>    <int> <chr>   <chr>       <dbl> <int>   <dbl>
#> 1            8   0.814    177       15 roc_auc binary      0.887    10  0.0103
#> 2            3  0.0402    151       10 roc_auc binary      0.885    10 0.00810
#> # … with 1 more variable: .config <chr>
```

There are other interim analysis techniques for discarding settings. For example, Krueger, Panknin, and Braun (2015) use traditional sequential analysis methods whereas Max Kuhn (2014) treats the data as a sports competition and uses the Bradley–Terry model (Bradley and Terry 1952) to measure the winning ability of parameter settings.

Chapter Summary

This chapter discussed the two main classes of grid search (regular and nonregular) that can be used for model tuning and demonstrated how to construct these grids, either manually or using the family of `grid_*()` functions. The `tune_grid()` function can evaluate these candidate sets of model parameters using resampling. The chapter also showed how to finalize a model, recipe, or workflow to update the parameter values for the final fit. Grid search can be computationally expensive, but thoughtful choices in the experimental design of such searches can make them tractable.

The data analysis code that will be reused in the next chapter is:

```
library(tidymodels)

data(cells)
cells <- cells %>% select(-case)

set.seed(1304)
cell_folds <- vfold_cv(cells)

roc_res <- metric_set(roc_auc)
```

## CHAPTER 14

Iterative Search

Chapter 13 demonstrated how grid search takes a predefined set of candidate values, evaluates them, then chooses the best settings. Iterative search methods pursue a different strategy. During the search process, they predict which values to test next.

When grid search is infeasible or inefficient, iterative methods are a sensible approach for optimizing tuning parameters.

This chapter outlines two search methods. First, we discuss *Bayesian optimization*, which uses a statistical model to predict better parameter settings. After that, the chapter describes a global search method called *simulated annealing*.

We use the same data on cell characteristics as the previous chapter for illustration but change the model. This chapter uses a support vector machine model because it provides nice two-dimensional visualizations of the search processes.

A Support Vector Machine Model

We once again use the cell segmentation data, described in Chapter 13, for modeling, with a support vector machine (SVM) model to demonstrate sequential tuning methods. See Kuhn and Johnson (2013) for more information on this model. The two tuning parameters to optimize are the SVM cost value and the radial basis function kernel parameter σ. Both parameters can have a profound effect on the model complexity and performance.

The SVM model uses a dot product and, for this reason, it is necessary to center and scale the predictors. Like the multilayer perceptron model, this model would benefit from the use of PCA feature extraction. However, we will not use this third tuning parameter in this chapter so that we can visualize the search process in two dimensions.

Along with the previously used objects (shown in the summary of Chapter 13), the tidymodels objects svm_rec, svm_spec, and svm_wflow define the model process:

```
library(tidymodels)
tidymodels_prefer()

svm_rec <-
  recipe(class ~ ., data = cells) %>%
  step_YeoJohnson(all_numeric_predictors()) %>%
  step_normalize(all_numeric_predictors())

svm_spec <-
  svm_rbf(cost = tune(), rbf_sigma = tune()) %>%
  set_engine("kernlab") %>%
  set_mode("classification")

svm_wflow <-
  workflow() %>%
  add_model(svm_spec) %>%
  add_recipe(svm_rec)
```

The default parameter ranges for the two tuning parameters cost and rbf_sigma are:

```
cost()
#> Cost (quantitative)
#> Transformer: log-2 [1e-100, Inf]
#> Range (transformed scale): [-10, 5]
rbf_sigma()
#> Radial Basis Function sigma (quantitative)
#> Transformer: log-10 [1e-100, Inf]
#> Range (transformed scale): [-10, 0]
```

For illustration, let's slightly change the kernel parameter range, to improve the visualizations of the search:

```
svm_param <-
  svm_wflow %>%
  extract_parameter_set_dials() %>%
  update(rbf_sigma = rbf_sigma(c(-7, -1)))
```

Before discussing specific details about iterative search and how it works, let's explore the relationship between the two SVM tuning parameters and the area under the ROC curve for this specific data set. We constructed a very large regular grid, composed of 2,500 candidate values, and evaluated the grid using resampling. This is obviously impractical in regular data analysis and tremendously inefficient. However, it elucidates the path that the search process should take and where the numerically optimal value(s) occur.

Figure 14-1 shows the results of evaluating this grid, with lighter color corresponding to higher (better) model performance. There is a large swath in the lower diagonal of the parameter space that is relatively flat with poor performance. A ridge of best performance occurs in the upper-right portion of the space. The black dot indicates the best settings. The transition from the plateau of poor results to the ridge of best performance is very sharp. There is also a sharp drop in the area under the ROC curve just to the right of the ridge.

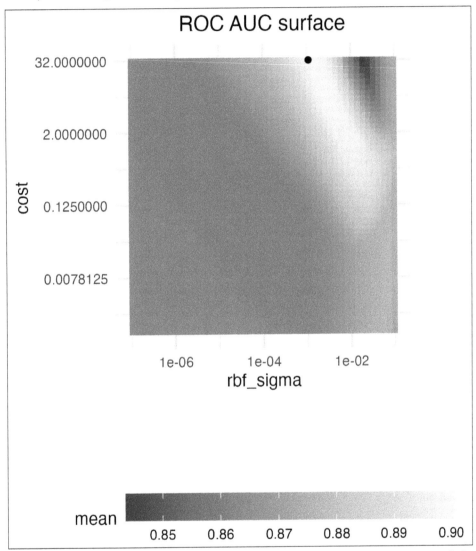

Figure 14-1. Heatmap of the mean area under the ROC curve for a high-density grid of tuning parameter values. The best point is a solid dot in the upper-right corner.

The following search procedures require at least some resampled performance statistics before proceeding. For this purpose, the following code creates a small regular grid that resides in the flat portion of the parameter space. The `tune_grid()` function resamples this grid:

```
set.seed(1401)
start_grid <-
  svm_param %>%
  update(
    cost = cost(c(-6, 1)),
    rbf_sigma = rbf_sigma(c(-6, -4))
  ) %>%
  grid_regular(levels = 2)

set.seed(1402)
svm_initial <-
  svm_wflow %>%
  tune_grid(resamples = cell_folds, grid = start_grid, metrics = roc_res)

collect_metrics(svm_initial)
#> # A tibble: 4 × 8
#>      cost rbf_sigma .metric .estimator  mean     n std_err .config
#>     <dbl>     <dbl> <chr>   <chr>      <dbl> <int>   <dbl> <chr>
#> 1 0.0156  0.000001 roc_auc binary     0.864    10 0.00864 Prepro...
#> 2 2       0.000001 roc_auc binary     0.863    10 0.00867 Prepro...
#> 3 0.0156  0.0001   roc_auc binary     0.863    10 0.00862 Prepro...
#> 4 2       0.0001   roc_auc binary     0.866    10 0.00855 Prepro...
```

This initial grid shows fairly equivalent results, with no individual point much better than any of the others. These results can be ingested by the iterative tuning functions discussed in the following sections to be used as initial values.

Bayesian Optimization

Bayesian optimization techniques analyze the current resampling results and create a predictive model to suggest tuning parameter values that have yet to be evaluated. The suggested parameter combination is then resampled. These results are then used in another predictive model that recommends more candidate values for testing, and so on. The process proceeds for a set number of iterations or until no further improvements occur. Shahriari et al. (2016) and Frazier (2018) are good introductions to Bayesian optimization.

When using Bayesian optimization, the primary concerns are how to create the model and how to select parameters recommended by that model. First, let's consider the technique most commonly used for Bayesian optimization, the Gaussian process model.

A Gaussian Process Model

Gaussian process (GP) (Schulz, Speekenbrink, and Krause 2018) models are well-known statistical techniques that have a history in spatial statistics (under the name of *kriging methods*). They can be derived in multiple ways, including as a Bayesian model; see Rasmussen and Williams (2006) for an excellent reference.

Mathematically, a GP is a collection of random variables whose joint probability distribution is multivariate Gaussian. In the context of our application, this is the collection of performance metrics for the tuning parameter candidate values. For the previous initial grid of four samples, the realizations of these four random variables were 0.8639, 0.8625, 0.8627, and 0.8659. These are assumed to be distributed as multivariate Gaussian. The inputs that define the independent variables/predictors for the GP model are the corresponding tuning parameter values (shown in Table 14-1).

Table 14-1. Resampling statistics used as the initial substrate to the Gaussian process model, where ROC is the outcome and both cost and rbf_sigma are predictors

ROC	cost	rbf_sigma
0.8639	0.01562	0.000001
0.8625	2.00000	0.000001
0.8627	0.01562	0.000100
0.8659	2.00000	0.000100

Gaussian process models are specified by their mean and covariance functions, although the latter has the most effect on the nature of the GP model. The covariance function is often parameterized in terms of the input values (denoted as x). As an example, a commonly used covariance function is the squared exponential[1] function:

$$\mathrm{cov}\left(\mathbf{x}_i, \mathbf{x}_j\right) = \exp\left(-\frac{1}{2}\left|\mathbf{x}_i - \mathbf{x}_j\right|^2\right) + \sigma_{ij}^2$$

where σ_{ij}^2 is a constant error variance term that is zero when $i = j$. This equation translates to:

> As the distance between two tuning parameter combinations increases, the covariance between the performance metrics increase exponentially.

The nature of the equation also implies that the variation of the outcome metric is minimized at the points that have already been observed (i.e., when $\left|\mathbf{x}_i - \mathbf{x}_j\right|^2$ is zero).

The nature of this covariance function allows the Gaussian process to represent highly nonlinear relationships between model performance and the tuning parameters even when only a small amount of data exists.

---

1 This equation is also the same as the *radial basis function* used in kernel methods, such as the SVM model that is currently being used. This is a coincidence; this covariance function is unrelated to the SVM tuning parameter that we are using.

However, fitting these models can be difficult in some cases, and the model becomes more computationally expensive as the number of tuning parameter combinations increases.

An important virtue of this model is that, since a full probability model is specified, the predictions for new inputs can reflect the entire distribution of the outcome. In other words, new performance statistics can be predicted in terms of both mean and variance.

Suppose that two new tuning parameters were under consideration. In Table 14-2, candidate A has a slightly better mean ROC value than candidate B (the current best is 0.8659). However, its variance is four-fold larger than B. Is this good or bad? Choosing option A is riskier but has potentially higher return. The increase in variance also reflects that this new value is farther from the existing data than B. The next section considers these aspects of GP predictions for Bayesian optimization in more detail.

Table 14-2. Two example tuning parameters considered for further sampling

Candidate	Mean	Variance
A	0.90	0.000400
B	0.89	0.000025

Bayesian optimization is an iterative process.

Based on the initial grid of four results, the GP model is fit, candidates are predicted, and a fifth tuning parameter combination is selected. We compute performance estimates for the new configuration, the GP is refit with the five existing results (and so on).

Acquisition Functions

Once the Gaussian process is fit to the current data, how is it used? Our goal is to choose the next tuning parameter combination that is most likely to have "better results" than the current best. One approach to do this is to create a large candidate set (perhaps using a space-filling design) and then make mean and variance predictions on each. Using this information, we choose the most advantageous tuning parameter value.

A class of objective functions, called *acquisition functions*, facilitate the trade-off between mean and variance. Recall that the predicted variance of the GP models are mostly driven by how far away they are from the existing data. The trade-off between the predicted mean and variance for new candidates is frequently viewed through the lens of exploration and exploitation:

Exploration

This biases the selection toward regions where there are fewer (if any) observed candidate models. This tends to give more weight to candidates with higher variance and focuses on finding new results.

Exploitation

This principally relies on the mean prediction to find the best (mean) value. It focuses on existing results.

To demonstrate, let's look at a toy example with a single parameter that has values between [0, 1] and the performance metric is R^2. The true function is shown in Figure 14-2, along with five candidate values that have existing results as points.

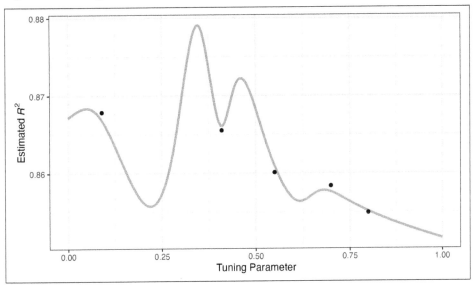

Figure 14-2. Hypothetical true performance profile over an arbitrary tuning parameter, with five estimated points.

For these data, the GP model fit is shown in Figure 14-3. The shaded region indicates the mean ± 1 standard error. The two vertical lines indicate two candidate points that are examined in more detail later.

The shaded confidence region demonstrates the squared exponential variance function; it becomes very large between points and converges to zero at the existing data points.

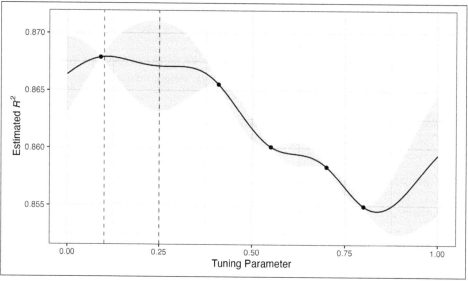

Figure 14-3. Estimated performance profile generated by the Gaussian process model. The shaded region shows one-standard-error bounds.

This nonlinear trend passes through each observed point, but the model is not perfect. There are no observed points near the true optimum setting and, in this region, the fit could be much better. Despite this, the GP model can effectively point us in the right direction.

From a pure exploitation standpoint, the best choice would select the parameter value that has the best mean prediction. Here, this would be a value of 0.106, just to the right of the existing best observed point at 0.09.

As a way to encourage exploration, a simple (but not often used) approach is to find the tuning parameter associated with the largest confidence interval. For example, by using a single standard deviation for the R^2 confidence bound, the next point to sample would be 0.236. This is slightly more into the region with no observed results. Increasing the number of standard deviations used in the upper bound would push the selection farther into empty regions.

One of the most commonly used acquisition functions is *expected improvement*. The notion of improvement requires a value for the current best results (unlike the confidence bound approach). Since the GP can describe a new candidate point using a distribution, we can weight the parts of the distribution that show improvement using the probability of the improvement occurring.

For example, consider two candidate parameter values of 0.10 and 0.25 (indicated by the vertical lines in Figure 14-3). Using the fitted GP model, their predicted R^2 distributions are shown in Figure 14-4 along with a reference line for the current best results.

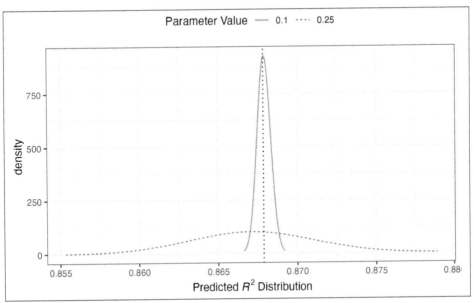

Figure 14-4. Predicted performance distributions for two sampled tuning parameter values.

When only considering the mean R^2 prediction, a parameter value of 0.10 is the better choice (see Table 14-3). The tuning parameter recommendation for 0.25 is, on average, predicted to be worse than the current best. However, since it has higher variance, it has more overall probability area above the current best. As a result, it has a larger expected improvement:

Table 14-3. Expected improvement for the two candidate tuning parameters

Parameter value	Mean	Std dev	Expected improvement
0.10	0.8679	0.0004317	0.000190
0.25	0.8671	0.0039301	0.001216

When expected improvement is computed across the range of the tuning parameter, the recommended point to sample is much closer to 0.25 than 0.10, as shown in Figure 14-5.

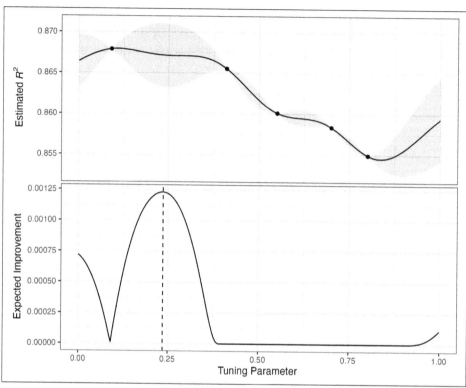

Figure 14-5. The estimated performance profile generated by the Gaussian process model (top panel) and the expected improvement (bottom panel). The vertical line indicates the point of maximum improvement.

Numerous acquisition functions have been proposed and discussed; in tidymodels, expected improvement is the default.

The tune_bayes() Function

To implement iterative search via Bayesian optimization, use the `tune_bayes()` function. Its syntax is very similar to `tune_grid()` but with several additional arguments:

`iter`

This is the maximum number of search iterations.

`initial`

This can be either an integer, an object produced using `tune_grid()`, or one of the racing functions. Using an integer specifies the size of a space-filling design that is sampled prior to the first GP model.

objective

> This is an argument for which acquisition function should be used. The tune package contains functions to pass here, such as `exp_improve()` or `conf_bound()`.

param_info *argument*

> In this case, this specifies the range of the parameters as well as any transformations that are used. These are used to define the search space. In situations where the default parameter objects are insufficient, `param_info` is used to override the defaults.

The `control` argument now uses the results of `control_bayes()`. Some helpful arguments there are:

no_improve

> This is an integer that will stop the search if improved parameters are not discovered within `no_improve` iterations.

uncertain

> This is also an integer (or `Inf`) that will take an *uncertainty sample* if there is no improvement within `uncertain` iterations. This will select the next candidate that has large variation. It has the effect of pure exploration since it does not consider the mean prediction.

verbose

> This is a logical that will print logging information as the search proceeds.

Let's use the first SVM results from the beginning of this chapter as the initial substrate for the Gaussian process model. Recall that, for this application, we want to maximize the area under the ROC curve. Our code is:

```
ctrl <- control_bayes(verbose = TRUE)

set.seed(1403)
svm_bo <-
  svm_wflow %>%
  tune_bayes(
    resamples = cell_folds,
    metrics = roc_res,
    initial = svm_initial,
    param_info = svm_param,
    iter = 25,
    control = ctrl
  )
```

The search process starts with an initial best value of 0.8659 for the area under the ROC curve. A Gaussian process model uses these four statistics to create a model. The large candidate set is automatically generated and scored using the expected improvement acquisition function. The first iteration failed to improve the outcome with an ROC value of 0.86315. After fitting another Gaussian process model with the new outcome value, the second iteration also failed to yield an improvement.

The log of the first two iterations, produced by the verbose option, was:

```
#> Optimizing roc_auc using the expected improvement
#>
#> ─ Iteration 1 ────────────────────────────────
#>
#> i Current best:      roc_auc=0.8659 (@iter 0)
#> i Gaussian process model
#> ✓ Gaussian process model
#> i Generating 5000 candidates
#> i Predicted candidates
#> i cost=0.386, rbf_sigma=0.000266
#> i Estimating performance
#> ✓ Estimating performance
#> ⊗ Newest results:    roc_auc=0.8631 (+/-0.00866)
#>
#> ─ Iteration 2 ────────────────────────────────
#>
#> i Current best:      roc_auc=0.8659 (@iter 0)
#> i Gaussian process model
#> ✓ Gaussian process model
#> i Generating 5000 candidates
#> i Predicted candidates
#> i cost=13.8, rbf_sigma=7.83e-07
#> i Estimating performance
#> ✓ Estimating performance
#> ⊗ Newest results:    roc_auc=0.8624 (+/-0.00865)
```

The search continues. There were a total of nine improvements in the outcome along the way at iterations 3, 4, 5, 6, 8, 13, 22, 23, and 24. The best result occurred at iteration 24 with an area under the ROC curve of 0.8986:

```
#> ─ Iteration 24 ───────────────────────────────
#>
#> i Current best:      roc_auc=0.8986 (@iter 23)
#> i Gaussian process model
#> ✓ Gaussian process model
#> i Generating 5000 candidates
#> i Predicted candidates
#> i cost=31.8, rbf_sigma=0.0016
#> i Estimating performance
#> ✓ Estimating performance
#> ♥ Newest results:    roc_auc=0.8986 (+/-0.00785)
```

The last step was:

```
#> ─ Iteration 25 ───────────────────────────────
#>
#> i Current best:      roc_auc=0.8986 (@iter 24)
#> i Gaussian process model
#> ✓ Gaussian process model
#> i Generating 5000 candidates
#> i Predicted candidates
#> i cost=20, rbf_sigma=0.00188
#> i Estimating performance
```

```
#> ✓ Estimating performance
#> ⊗ Newest results:    roc_auc=0.8982 (+/-0.00781)
```

The functions that are used to interrogate the results are the same as those used for grid search (e.g., collect_metrics(), etc.). For example:

```
show_best(svm_bo)
#> # A tibble: 5 × 9
#>     cost rbf_sigma .metric .estimator  mean     n std_err .config .iter
#>    <dbl>     <dbl> <chr>   <chr>      <dbl> <int>   <dbl> <chr>   <int>
#> 1  31.8   0.00160 roc_auc binary     0.899    10 0.00785 Iter24     24
#> 2  30.8   0.00191 roc_auc binary     0.899    10 0.00791 Iter23     23
#> 3  31.4   0.00166 roc_auc binary     0.899    10 0.00784 Iter22     22
#> 4  31.8   0.00153 roc_auc binary     0.899    10 0.00783 Iter13     13
#> 5  30.8   0.00163 roc_auc binary     0.899    10 0.00782 Iter15     15
```

The autoplot() function has several options for iterative search methods. Figure 14-6 shows how the outcome changed over the search by using autoplot(svm_bo, type = "performance").

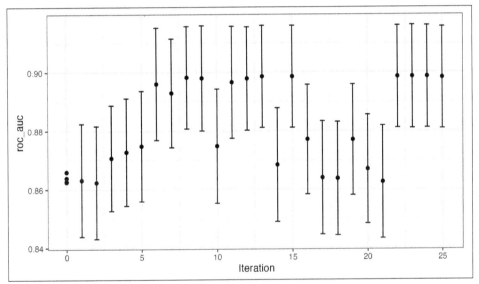

Figure 14-6. *The progress of the Bayesian optimization produced when the* autoplot() *method is used with* type = "performance".

An additional type of plot uses type = "parameters" that shows the parameter values over iterations.

Figure 14-7 shows the surfaces of the mean, variance, and expected improvement surfaces estimated by the GP after 11 iterations. The panel on the right shows a ridge of best estimated improvement along the right side of the candidate space.

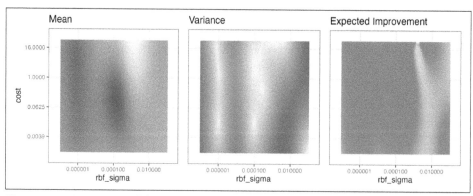

Figure 14-7. Heatmaps of the predicted mean RMSE (left), variance of RMSE (middle), and the expected improvement (right) after 11 search iterations.

Figure 14-8 shows the search process at three different points in the optimization.

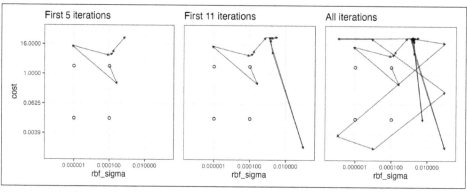

Figure 14-8. The Bayesian optimization search path after 5, 11, and 25 iterations.

The first five iterations initially moved in a poor direction but quickly moved closer to better results. The middle panel shows the first eleven iterations where the process investigates the region of true optimal results with a short foray to the bottom right boundary of the candidate space. The remaining iterations shown in the panel on the left switch between the region of best results and the far borders of the search space.

While the best tuning parameter combination is on the boundary of the parameter space, Bayesian optimization will often choose new points on other sides of the boundary. While we can adjust the ratio of exploration and exploitation, the search tends to sample boundary points early on.

Finally, if the user interrupts the `tune_bayes()` computations, the function returns the current results (instead of resulting in an error).

If the search is seeded with an initial grid, a space-filling design would probably be a better choice than a regular design. It samples more unique values of the parameter space and would improve the predictions of the standard deviation in the early iterations.

Simulated Annealing

Simulated annealing (SA) (Kirkpatrick, Gelatt, and Vecchi 1983; Van Laarhoven and Aarts 1987) is a general nonlinear search routine inspired by the process by which metal cools. It is a global search method that can effectively navigate many different types of search landscapes, including discontinuous functions. Unlike most gradient-based optimization routines, simulated annealing can reassess previous solutions.

Simulated Annealing Search Process

The process of using simulated annealing starts with an initial value and embarks on a controlled random walk through the parameter space. Each new candidate parameter value is a small perturbation of the previous value that keeps the new point within a local neighborhood.

The candidate point is resampled to obtain its corresponding performance value. If this achieves better results than the previous parameters, it is accepted as the new best and the process continues. If the results are worse than the previous value, the search procedure may still use this parameter to define further steps. This depends on two factors. First, the likelihood of accepting a bad result decreases as performance becomes worse. In other words, a slightly worse result has a better chance of acceptance than one with a large drop in performance. The other factor is the number of search iterations. Simulated annealing wants to accept fewer suboptimal values as the search proceeds. From these two factors, the *acceptance probability* for a bad result can be formalized as:

$$\Pr\left[\text{accept suboptimal parameters at iteration } i\right] = \exp\left(c \times D_i \times i\right)$$

where i is the iteration number, c is a user-specified constant, and D_i is the percent difference between the old and new values (where negative values imply worse results). For a bad result, we determine the acceptance probability and compare it to a random uniform number. If the random number is greater than the probability value, the search discards the current parameters and the next iteration creates its candidate value in the neighborhood of the previous value. Otherwise, the next iteration forms the next set of parameters based on the current (suboptimal) values.

The acceptance probabilities of simulated annealing allow the search to proceed in the wrong direction, at least for the short term, with the potential to find a much better region of the parameter space in the long run.

How are the acceptance probabilities influenced? The heatmap in Figure 14-9 shows how the acceptance probability can change over iterations, performance, and the user-specified coefficient.

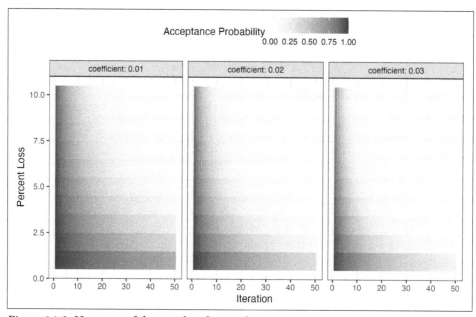

Figure 14-9. Heatmap of the simulated annealing acceptance probabilities for different coefficient values.

The user can adjust the coefficients to find a probability profile that suits their needs. In finetune::control_sim_anneal(), the default for this cooling_coef argument is 0.02. Decreasing this coefficient will encourage the search to be more forgiving of poor results.

This process continues for a set amount of iterations but can halt if no globally best results occur within a predetermined number of iterations. However, it can be very helpful to set a *restart threshold*. If there are a string of failures, this feature revisits the last globally best parameter settings and starts anew.

The main important detail is to define how to perturb the tuning parameters from iteration to iteration. There are a variety of methods in the literature for this. We follow the method given in Bohachevsky, Johnson, and Stein (1986) called *generalized simulated annealing*. For continuous tuning parameters, we define a small radius to specify the local "neighborhood." For example, suppose there are two tuning parameters and each is bounded by zero and one. The simulated annealing process generates random values on the surrounding radius and randomly chooses one to be the current candidate value.

In our implementation, the neighborhood is determined by scaling the current candidate to be between zero and one based on the range of the parameter object, so radius values between 0.05 and 0.15 seem reasonable. For these values, the fastest that the search could go from one side of the parameter space to the other is about 10 iterations. The size of the radius controls how quickly the search explores the parameter space. In our implementation, a range of radii is specified so different magnitudes of "local" define the new candidate values.

To illustrate, we'll use the two main glmnet tuning parameters:

- The amount of total regularization (penalty). The default range for this parameter is 10^{-10} to 10^0. It is typical to use a log (base-10) transformation for this parameter.

- The proportion of the lasso penalty (mixture). This is bounded at zero and one with no transformation.

The process starts with initial values of penalty = 0.025 and mixture = 0.050. Using a radius that randomly fluctuates between 0.050 and 0.015, the data are appropriately scaled, random values are generated on radii around the initial point, and then one is randomly chosen as the candidate. For illustration, we will assume that all candidate values are improvements. Using the new value, a set of new random neighbors are generated, one is chosen, and so on. Figure 14-10 shows six iterations as the search proceeds toward the upper left corner.

Note that, during some iterations, the candidate sets along the radius exclude points outside of the parameter boundaries. Also, our implementation biases the choice of the next tuning parameter configurations *away* from new values that are very similar to previous configurations.

For nonnumeric parameters, we assign a probability for how often the parameter value changes.

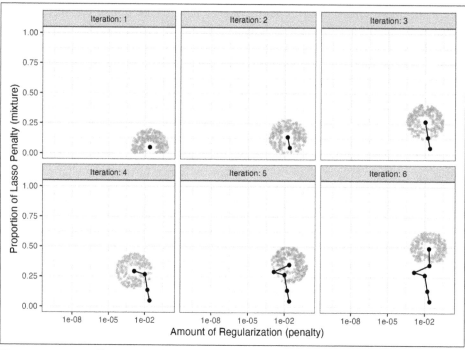

Figure 14-10. An illustration of how simulated annealing determines what is the local neighborhood for two numeric tuning parameters. The clouds of points show possible next values where one would be selected at random.

The tune_sim_anneal() Function

To implement iterative search via simulated annealing, use the tune_sim_anneal() function. The syntax for this function is nearly identical to tune_bayes(). There are no options for acquisition functions or uncertainty sampling. The control_sim_anneal() function has some details that define the local neighborhood and the cooling schedule:

- no_improve, for simulated annealing, is an integer that will stop the search if no global best or improved results are discovered within no_improve iterations. Accepted suboptimal or discarded parameters count as "no improvement."

- restart is the number of iterations with no new best results before starting from the previous best results.

- radius is a numeric vector on (0, 1) that defines the minimum and maximum radius of the local neighborhood around the initial point.

- flip is a probability value that defines the chances of altering the value of categorical or integer parameters.

- cooling_coef is the c coefficient in $\exp(c \times D_i \times i)$ that modulates how quickly the acceptance probability decreases over iterations. Larger values of cooling_coef decrease the probability of accepting a suboptimal parameter setting.

For the cell segmentation data, the syntax is very consistent with the previously used functions:

```
ctrl_sa <- control_sim_anneal(verbose = TRUE, no_improve = 10L)

set.seed(1404)
svm_sa <-
  svm_wflow %>%
  tune_sim_anneal(
    resamples = cell_folds,
    metrics = roc_res,
    initial = svm_initial,
    param_info = svm_param,
    iter = 50,
    control = ctrl_sa
  )
```

The simulated annealing process discovered new global optimums at 4 different iterations. The earliest improvement was at Iteration 5 and the final optimum occured at Iteration 27. The best overall results occured at Iteration 27 with a mean area under the ROC curve of 0.8985 (compared to an initial best of 0.8659). There were 4 restarts at Iterations 13, 21, 35, and 43 as well as 12 discarded candidates during the process.

The verbose option prints details of the search process. The output for the first five iterations was:

```
#> Optimizing roc_auc
#> Initial best: 0.86594
#>   1 ○ accept suboptimal  roc_auc=0.86351  (+/-0.008642)
#>   2 ○ accept suboptimal  roc_auc=0.86233  (+/-0.008657)
#>   3 + better suboptimal  roc_auc=0.86233  (+/-0.008661)
#>   4 + better suboptimal  roc_auc=0.86492  (+/-0.008504)
#>   5 ♥ new best           roc_auc=0.87247  (+/-0.008232)
```

The output for the last 10 iterations was:

```
#> 40 ○ accept suboptimal  roc_auc=0.89606  (+/-0.008203)
#> 41 - discard suboptimal roc_auc=0.87556  (+/-0.009272)
#> 42 - discard suboptimal roc_auc=0.87198  (+/-0.009301)
#> 43 ✖ restart from best   roc_auc=0.89801  (+/-0.008224)
#> 44 ○ accept suboptimal  roc_auc=0.89006  (+/-0.008789)
#> 45 + better suboptimal  roc_auc=0.89781  (+/-0.008104)
#> 46 ○ accept suboptimal  roc_auc=0.89563  (+/-0.008601)
#> 47 - discard suboptimal roc_auc=0.88527  (+/-0.008766)
#> 48 ○ accept suboptimal  roc_auc=0.8922   (+/-0.008891)
#> 49 - discard suboptimal roc_auc=0.87691  (+/-0.008352)
#> 50 ○ accept suboptimal  roc_auc=0.88803  (+/-0.008728)
```

As with the other tune_*() functions, the corresponding autoplot() function produces visual assessments of the results. Using autoplot(svm_sa, type =

`"performance"`) shows the performance over iterations (Figure 14-11) while `autoplot(svm_sa, type = "parameters")` plots performance versus specific tuning parameter values (Figure 14-12).

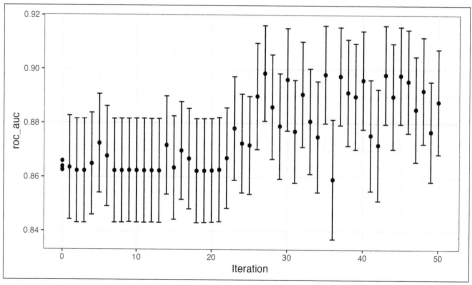

Figure 14-11. Progress of the simulated annealing process shown when the `autoplot()` method is used with `type = "performance"`.

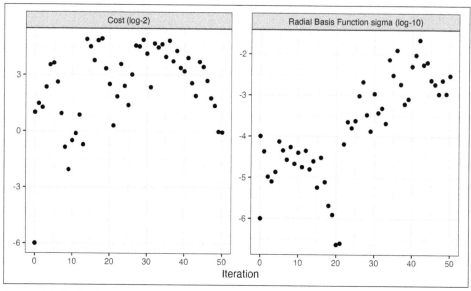

Figure 14-12. Performance versus tuning parameter values when the `autoplot()` method is used with `type = "parameters"`.

Like `tune_bayes()`, manually stopping execution will return the completed iterations.

A visualization of the search path helps to understand where the search process did well and where it went astray. Figure 14-13 illustrates several phases of the optimization; these are separated by a restart of the process at the last best results.

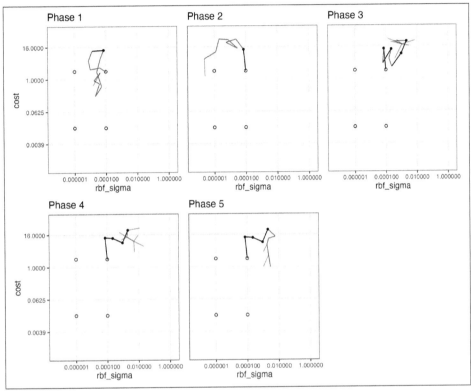

Figure 14-13. A visualization of different phases of the simulated annealing search.

In the first phase, the search initially finds two new global optima (shown with the solid points). From these, several settings are immediately discarded (light gray lines) while others are suboptimal but acceptable. After a set number of failures, it restarts at the last solid point. The other phases show a slow improvement in global optima with many discarded settings along the way. The process eventually finds its way to the region of optimal results as it exhausts the total number of allowed iterations.

Chapter Summary

This chapter described two iterative search methods for optimizing tuning parameters. Bayes optimization uses a predictive model trained on existing resampling results to suggest tuning parameter values, while simulated annealing walks through the hyperparameter space to find good values. Both can be effective at finding good values alone or as a follow-up method used after an initial grid search to further finetune performance.

## CHAPTER 15

Screening Many Models

We introduced workflow sets in Chapter 7 and demonstrated how to use them with resampled data sets in Chapter 11. In this chapter, we discuss these sets of multiple modeling workflows in more detail and describe a use case where they can be helpful.

For projects with new data sets that have not yet been well understood, a data practitioner may need to screen many combinations of models and preprocessors. It is common to have little or no a priori knowledge about which method will work best with a novel data set.

 A good strategy is to spend some initial effort trying a variety of modeling approaches, determine what works best, then invest additional time tweaking/optimizing a small set of models.

Workflow sets provide a user interface to create and manage this process. We'll also demonstrate how to evaluate these models efficiently using the racing methods discussed later in this chapter.

Modeling Concrete Mixture Strength

To demonstrate how to screen multiple model workflows, we will use the concrete mixture data from *Applied Predictive Modeling* (Kuhn and Johnson 2013) as an example. Chapter 10 of that book demonstrated models to predict the compressive strength of concrete mixtures using the ingredients as predictors. A wide variety of models were evaluated with different predictor sets and preprocessing needs. How can workflow sets make such a process of large-scale testing for models easier?

First, let's define the data splitting and resampling schemes:

```
library(tidymodels)
tidymodels_prefer()
data(concrete, package = "modeldata")
glimpse(concrete)
#> Rows: 1,030
#> Columns: 9
#> $ cement             <dbl> 540.0, 540.0, 332.5, 332.5, 198.6, 266.0, …
#> $ blast_furnace_slag <dbl> 0.0, 0.0, 142.5, 142.5, 132.4, 114.0, 95.0, …
#> $ fly_ash            <dbl> 0, 0, 0, 0, 0, 0, 0, 0, 0, 0, 0, 0, 0, …
#> $ water              <dbl> 162, 162, 228, 228, 192, 228, 228, 228, 228, …
#> $ superplasticizer   <dbl> 2.5, 2.5, 0.0, 0.0, 0.0, 0.0, 0.0, 0.0, 0.0, …
#> $ coarse_aggregate   <dbl> 1040.0, 1055.0, 932.0, 932.0, 978.4, 932.0, …
#> $ fine_aggregate     <dbl> 676.0, 676.0, 594.0, 594.0, 825.5, 670.0, …
#> $ age                <int> 28, 28, 270, 365, 360, 90, 365, 28, 28, 28, …
#> $ compressive_strength <dbl> 79.99, 61.89, 40.27, 41.05, 44.30, 47.03, …
```

The compressive_strength column is the outcome. The age predictor tells us the age of the concrete sample at testing in days (concrete strengthens over time) and the rest of the predictors like cement and water are concrete components in units of kilograms per cubic meter.

 For some cases in this data set the same concrete formula was tested multiple times. We'd rather not include these replicate mixtures as individual data points since they might be distributed across both the training and test set. Doing so might artificially inflate our performance estimates.

To address this, we will use the mean compressive strength per concrete mixture for modeling:

```
concrete <-
  concrete %>%
  group_by(across(-compressive_strength)) %>%
  summarize(compressive_strength = mean(compressive_strength),
            .groups = "drop")
nrow(concrete)
#> [1] 992
```

Let's split the data using the default 3:1 ratio of training-to-test and resample the training set using five repeats of 10-fold cross-validation:

```
set.seed(1501)
concrete_split <- initial_split(concrete, strata = compressive_strength)
concrete_train <- training(concrete_split)
concrete_test  <- testing(concrete_split)

set.seed(1502)
concrete_folds <-
  vfold_cv(concrete_train, strata = compressive_strength, repeats = 5)
```

Some models (notably neural networks, *K*-nearest neighbors, and support vector machines) require predictors that have been centered and scaled, so some model workflows will require recipes with these preprocessing steps. For other models, a traditional response surface design model expansion (i.e., quadratic and two-way interactions) is a good idea. For these purposes, we create two recipes:

```
normalized_rec <-
   recipe(compressive_strength ~ ., data = concrete_train) %>%
   step_normalize(all_predictors())

poly_recipe <-
   normalized_rec %>%
   step_poly(all_predictors()) %>%
   step_interact(~ all_predictors():all_predictors())
```

For the models, we use the the parsnip add-in to create a set of model specifications:

```
library(rules)
library(baguette)

linear_reg_spec <-
   linear_reg(penalty = tune(), mixture = tune()) %>%
   set_engine("glmnet")

nnet_spec <-
   mlp(hidden_units = tune(), penalty = tune(), epochs = tune()) %>%
   set_engine("nnet", MaxNWts = 2600) %>%
   set_mode("regression")

mars_spec <-
   mars(prod_degree = tune()) %>%   #<- use GCV to choose terms
   set_engine("earth") %>%
   set_mode("regression")

svm_r_spec <-
   svm_rbf(cost = tune(), rbf_sigma = tune()) %>%
   set_engine("kernlab") %>%
   set_mode("regression")

svm_p_spec <-
   svm_poly(cost = tune(), degree = tune()) %>%
   set_engine("kernlab") %>%
   set_mode("regression")

knn_spec <-
   nearest_neighbor(neighbors = tune(), dist_power = tune(), weight_func = tune()) %>%
   set_engine("kknn") %>%
   set_mode("regression")

cart_spec <-
   decision_tree(cost_complexity = tune(), min_n = tune()) %>%
   set_engine("rpart") %>%
   set_mode("regression")

bag_cart_spec <-
   bag_tree() %>%
   set_engine("rpart", times = 50L) %>%
   set_mode("regression")
```

```
rf_spec <-
  rand_forest(mtry = tune(), min_n = tune(), trees = 1000) %>%
  set_engine("ranger") %>%
  set_mode("regression")

xgb_spec <-
  boost_tree(tree_depth = tune(), learn_rate = tune(), loss_reduction = tune(),
             min_n = tune(), sample_size = tune(), trees = tune()) %>%
  set_engine("xgboost") %>%
  set_mode("regression")

cubist_spec <-
  cubist_rules(committees = tune(), neighbors = tune()) %>%
  set_engine("Cubist")
```

The analysis in Kuhn and Johnson (2013) specifies that the neural network should have up to 27 hidden units in the layer. The `extract_parameter_set_dials()` function extracts the parameter set, which we modify to have the correct parameter range:

```
nnet_param <-
  nnet_spec %>%
  extract_parameter_set_dials() %>%
  update(hidden_units = hidden_units(c(1, 27)))
```

How can we match these models to their recipes, tune them, then evaluate their performance efficiently? A workflow set offers a solution.

Creating the Workflow Set

Workflow sets take named lists of preprocessors and model specifications and combine them into an object containing multiple workflows. There are three possible kinds of preprocessors:

- A standard R formula
- A recipe object (prior to estimation/prepping)
- A dplyr-style selector to choose the outcome and predictors

As a first workflow set example, let's combine the recipe that only standardizes the predictors to the nonlinear models that require the predictors to be in the same units:

```
normalized <-
  workflow_set(
    preproc = list(normalized = normalized_rec),
    models = list(SVM_radial = svm_r_spec, SVM_poly = svm_p_spec,
                  KNN = knn_spec, neural_network = nnet_spec)
  )
normalized
#> # A workflow set/tibble: 4 × 4
#>   wflow_id              info             option    result
#>   <chr>                 <list>           <list>    <list>
#> 1 normalized_SVM_radial <tibble [1 × 4]> <opts[0]> <list [0]>
#> 2 normalized_SVM_poly   <tibble [1 × 4]> <opts[0]> <list [0]>
```

```
#> 3 normalized_KNN              <tibble [1 × 4]> <opts[0]> <list [0]>
#> 4 normalized_neural_network <tibble [1 × 4]> <opts[0]> <list [0]>
```

Since there is only a single preprocessor, this function creates a set of workflows with this value. If the preprocessor contained more than one entry, the function would create all combinations of preprocessors and models.

The wflow_id column is automatically created but can be modified using a call to mutate(). The info column contains a tibble with some identifiers and the workflow object. The workflow can be extracted:

```
normalized %>% extract_workflow(id = "normalized_KNN")
#> ══ Workflow ═══════════════════════════════════════
#> Preprocessor: Recipe
#> Model: nearest_neighbor()
#>
#> ── Preprocessor ───────────────────────────────────
#> 1 Recipe Step
#>
#> • step_normalize()
#>
#> ── Model ──────────────────────────────────────────
#> K-Nearest Neighbor Model Specification (regression)
#>
#> Main Arguments:
#>   neighbors = tune()
#>   weight_func = tune()
#>   dist_power = tune()
#>
#> Computational engine: kknn
```

The option column is a placeholder for any arguments to use when we evaluate the workflow. For example, to add the neural network parameter object:

```
normalized <-
  normalized %>%
  option_add(param_info = nnet_param, id = "normalized_neural_network")
normalized
#> # A workflow set/tibble: 4 × 4
#>   wflow_id                   info            option    result
#>   <chr>                      <list>          <list>    <list>
#> 1 normalized_SVM_radial      <tibble [1 × 4]> <opts[0]> <list [0]>
#> 2 normalized_SVM_poly        <tibble [1 × 4]> <opts[0]> <list [0]>
#> 3 normalized_KNN             <tibble [1 × 4]> <opts[0]> <list [0]>
#> 4 normalized_neural_network <tibble [1 × 4]> <opts[1]> <list [0]>
```

When a function from the tune or finetune package is used to tune (or resample) the workflow, this argument will be used.

The result column is a placeholder for the output of the tuning or resampling functions.

For the other nonlinear models, let's create another workflow set that uses dplyr selectors for the outcome and predictors:

```
model_vars <-
  workflow_variables(outcomes = compressive_strength,
                     predictors = everything())

no_pre_proc <-
  workflow_set(
    preproc = list(simple = model_vars),
    models = list(MARS = mars_spec,
                  CART = cart_spec,
                  CART_bagged = bag_cart_spec,
                  RF = rf_spec,
                  boosting = xgb_spec,
                  Cubist = cubist_spec)
  )
no_pre_proc
#> # A workflow set/tibble: 6 × 4
#>   wflow_id            info             option     result
#>   <chr>               <list>           <list>     <list>
#> 1 simple_MARS         <tibble [1 × 4]> <opts[0]>  <list [0]>
#> 2 simple_CART         <tibble [1 × 4]> <opts[0]>  <list [0]>
#> 3 simple_CART_bagged  <tibble [1 × 4]> <opts[0]>  <list [0]>
#> 4 simple_RF           <tibble [1 × 4]> <opts[0]>  <list [0]>
#> 5 simple_boosting     <tibble [1 × 4]> <opts[0]>  <list [0]>
#> 6 simple_Cubist       <tibble [1 × 4]> <opts[0]>  <list [0]>
```

Finally, we assemble the set that uses nonlinear terms and interactions with the appropriate models:

```
with_features <-
  workflow_set(
    preproc = list(full_quad = poly_recipe),
    models = list(linear_reg = linear_reg_spec, KNN = knn_spec)
  )
```

These objects are tibbles with the extra class of workflow_set. Row binding does not affect the state of the sets, and the result is itself a workflow set:

```
all_workflows <-
  bind_rows(no_pre_proc, normalized, with_features) %>%
  # Make the workflow IDs a little more simple:
  mutate(wflow_id = gsub("(simple_)|(normalized_)", "", wflow_id))
all_workflows
#> # A workflow set/tibble: 12 × 4
#>   wflow_id     info             option     result
#>   <chr>        <list>           <list>     <list>
#> 1 MARS         <tibble [1 × 4]> <opts[0]>  <list [0]>
#> 2 CART         <tibble [1 × 4]> <opts[0]>  <list [0]>
#> 3 CART_bagged  <tibble [1 × 4]> <opts[0]>  <list [0]>
#> 4 RF           <tibble [1 × 4]> <opts[0]>  <list [0]>
#> 5 boosting     <tibble [1 × 4]> <opts[0]>  <list [0]>
#> 6 Cubist       <tibble [1 × 4]> <opts[0]>  <list [0]>
#> # … with 6 more rows
```

Tuning and Evaluating the Models

Almost all of the members of `all_workflows` contain tuning parameters. To evaluate their performance, we can use the standard tuning or resampling functions (e.g., `tune_grid()` and so on). The `workflow_map()` function will apply the same function to all of the workflows in the set; the default is `tune_grid()`.

For this example, grid search is applied to each workflow using up to 25 different parameter candidates. There is a set of common options to use with each execution of `tune_grid()`. For example, in the following code we will use the same resampling and control objects for each workflow, along with a grid size of 25. The `workflow_map()` function has an additional argument called `seed`, which is used to ensure that each execution of `tune_grid()` consumes the same random numbers:

```
grid_ctrl <-
  control_grid(
    save_pred = TRUE,
    parallel_over = "everything",
    save_workflow = TRUE
  )

grid_results <-
  all_workflows %>%
  workflow_map(
    seed = 1503,
    resamples = concrete_folds,
    grid = 25,
    control = grid_ctrl
  )
```

The results show that the `option` and `result` columns have been updated:

```
grid_ctrl <-
  control_grid(
    save_pred = TRUE,
    parallel_over = "everything",
    save_workflow = TRUE
  )

full_results_time <-
  system.time(
    grid_results <-
      all_workflows %>%
      workflow_map(seed = 1503, resamples = concrete_folds, grid = 25,
                   control = grid_ctrl, verbose = TRUE)
  )
#> i  1 of 12 tuning:     MARS
#> ✓  1 of 12 tuning:     MARS (2.7s)
#> i  2 of 12 tuning:     CART
#> ✓  2 of 12 tuning:     CART (27.6s)
#> i     No tuning parameters. `fit_resamples()` will be attempted
#> i  3 of 12 resampling: CART_bagged
#> ✓  3 of 12 resampling: CART_bagged (18.5s)
#> i  4 of 12 tuning:     RF
#> i Creating preprocessing data to finalize unknown parameter: mtry
```

```
#> ✓  4 of 12 tuning:    RF (1m 9.2s)
#> i  5 of 12 tuning:    boosting
#> ✓  5 of 12 tuning:    boosting (2m 4.1s)
#> i  6 of 12 tuning:    Cubist
#> ✓  6 of 12 tuning:    Cubist (2m 0.7s)
#> i  7 of 12 tuning:    SVM_radial
#> ✓  7 of 12 tuning:    SVM_radial (40.2s)
#> i  8 of 12 tuning:    SVM_poly
#> ✓  8 of 12 tuning:    SVM_poly (7m 46.4s)
#> i  9 of 12 tuning:    KNN
#> ✓  9 of 12 tuning:    KNN (43.2s)
#> i 10 of 12 tuning:    neural_network
#> ✓ 10 of 12 tuning:    neural_network (1m 22s)
#> i 11 of 12 tuning:    full_quad_linear_reg
#> ✓ 11 of 12 tuning:    full_quad_linear_reg (57.9s)
#> i 12 of 12 tuning:    full_quad_KNN
#> ✓ 12 of 12 tuning:    full_quad_KNN (2m 59.8s)

num_grid_models <- nrow(collect_metrics(grid_results, summarize = FALSE))
```

What do our `grid_results` look like?

```
grid_results
#> # A workflow set/tibble: 12 × 4
#>   wflow_id    info            option    result
#>   <chr>       <list>          <list>    <list>
#> 1 MARS        <tibble [1 × 4]> <opts[3]> <tune[+]>
#> 2 CART        <tibble [1 × 4]> <opts[3]> <tune[+]>
#> 3 CART_bagged <tibble [1 × 4]> <opts[3]> <rsmp[+]>
#> 4 RF          <tibble [1 × 4]> <opts[3]> <tune[+]>
#> 5 boosting    <tibble [1 × 4]> <opts[3]> <tune[+]>
#> 6 Cubist      <tibble [1 × 4]> <opts[3]> <tune[+]>
#> # … with 6 more rows
```

The `option` column now contains all of the options that we used in the `workflow_map()` call. This makes our results reproducible. In the `result` columns, the `"tune[+]"` and `"rsmp[+]"` notations mean that the object had no issues. A value such as `"tune[x]"` occurs if all of the models failed for some reason.

There are a few convenience functions for examining results such as `grid_results`. The `rank_results()` function will order the models by some performance metric. By default, it uses the first metric in the metric set (RMSE in this instance). Let's `filter()` to look only at RMSE:

```
grid_results %>%
  rank_results() %>%
  filter(.metric == "rmse") %>%
  select(model, .config, rmse = mean, rank)
#> # A tibble: 252 × 4
#>   model      .config                rmse  rank
#>   <chr>      <chr>                 <dbl> <int>
#> 1 boost_tree Preprocessor1_Model04  4.25     1
#> 2 boost_tree Preprocessor1_Model06  4.29     2
#> 3 boost_tree Preprocessor1_Model13  4.31     3
#> 4 boost_tree Preprocessor1_Model14  4.39     4
#> 5 boost_tree Preprocessor1_Model16  4.46     5
```

```
#> 6 boost_tree Preprocessor1_Model03  4.47      6
#> # … with 246 more rows
```

Also by default, the function ranks all of the candidate sets; that's why the same model can show up multiple times in the output. An option, called select_best, can be used to rank the models using their best tuning parameter combination.

The autoplot() method plots the rankings; it also has a select_best argument. The plot in Figure 15-1 visualizes the best results for each model and is generated with:

```
autoplot(
    grid_results,
    rank_metric = "rmse",   # <- how to order models
    metric = "rmse",        # <- which metric to visualize
    select_best = TRUE      # <- one point per workflow
) +
    geom_text(aes(y = mean - 1/2, label = wflow_id), angle = 90, hjust = 1) +
    lims(y = c(3.5, 9.5)) +
    theme(legend.position = "none")
```

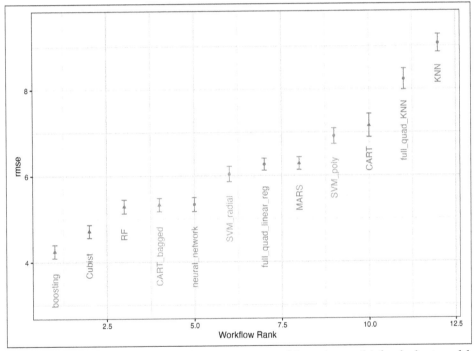

Figure 15-1. Estimated RMSE (and approximate confidence intervals) for the best model configuration in each workflow.

In case you want to see the tuning parameter results for a specific model, like Figure 15-2, the id argument can take a single value from the wflow_id column for which model to plot:

```
autoplot(grid_results, id = "Cubist", metric = "rmse")
```

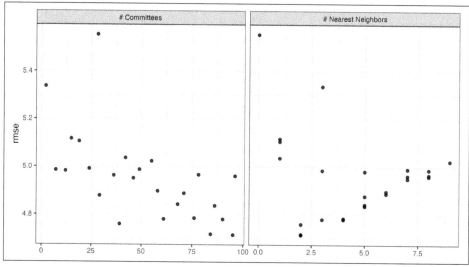

Figure 15-2. The `autoplot()` *results for the Cubist model contained in the workflow set.*

There are also methods for `collect_predictions()` and `collect_metrics()`.

The example model screening with our concrete mixture data fits a total of 25,200 models. Using two workers in parallel, the estimation process took 1.9 hours to complete.

Efficiently Screening Models

One effective method for screening a large set of models efficiently is to use the racing approach described in Chapter 13. With a workflow set, we can use the `workflow_map()` function for this racing approach. Recall that after we pipe in our workflow set, the argument we use is the function to apply to the workflows; in this case, we can use a value of `"tune_race_anova"`. We also pass an appropriate control object; otherwise, the options would be the same as the code in the previous section:

```
library(finetune)

race_ctrl <-
   control_race(
      save_pred = TRUE,
      parallel_over = "everything",
      save_workflow = TRUE
   )

race_results <-
   all_workflows %>%
   workflow_map(
      "tune_race_anova",
      seed = 1503,
      resamples = concrete_folds,
      grid = 25,
      control = race_ctrl
   )
```

The new object looks very similar, although the elements of the `result` column show a value of `"race[+]"`, indicating a different type of object:

```
race_results
#> # A workflow set/tibble: 12 × 4
#>   wflow_id   info            option    result
#>   <chr>      <list>          <list>    <list>
#> 1 MARS       <tibble [1 × 4]> <opts[3]> <race[+]>
#> 2 CART       <tibble [1 × 4]> <opts[3]> <race[+]>
#> 3 CART_bagged <tibble [1 × 4]> <opts[3]> <rsmp[+]>
#> 4 RF         <tibble [1 × 4]> <opts[3]> <race[+]>
#> 5 boosting   <tibble [1 × 4]> <opts[3]> <race[+]>
#> 6 Cubist     <tibble [1 × 4]> <opts[3]> <race[+]>
#> # … with 6 more rows
```

The same helpful functions are available for this object to interrogate the results and, in fact, the basic `autoplot()` method shown in Figure 15-3[1] produces trends similar to Figure 15-2. This is produced by:

```
autoplot(
   race_results,
   rank_metric = "rmse",
   metric = "rmse",
   select_best = TRUE
) +
   geom_text(aes(y = mean - 1/2, label = wflow_id), angle = 90, hjust = 1) +
   lims(y = c(3.0, 9.5)) +
   theme(legend.position = "none")
```

Overall, the racing approach estimated a total of 4,652 models, 18.46% of the full set of 25,200 models in the full grid. As a result, the racing approach was 4.5-fold faster.

---

1 As of February 2022, we see slightly different performance metrics for the neural network when trained using macOS on ARM architecture (Apple M1 chip) compared to Intel architecture.

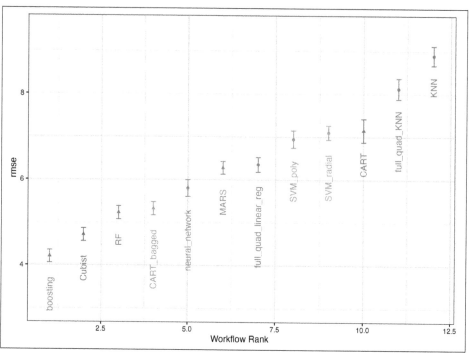

Figure 15-3. Estimated RMSE (and approximate confidence intervals) for the best model configuration in each workflow in the racing results.

Did we get similar results? For both objects, we rank the results, merge them, and plot them against one another in Figure 15-4:

```
matched_results <-
  rank_results(race_results, select_best = TRUE) %>%
  select(wflow_id, .metric, race = mean, config_race = .config) %>%
  inner_join(
    rank_results(grid_results, select_best = TRUE) %>%
      select(wflow_id, .metric, complete = mean,
             config_complete = .config, model),
    by = c("wflow_id", ".metric"),
  ) %>%
  filter(.metric == "rmse")

library(ggrepel)

matched_results %>%
  ggplot(aes(x = complete, y = race)) +
  geom_abline(lty = 3) +
  geom_point() +
  geom_text_repel(aes(label = model)) +
  coord_obs_pred() +
  labs(x = "Complete Grid RMSE", y = "Racing RMSE")
```

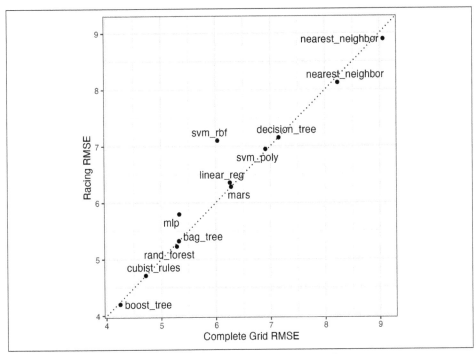

Figure 15-4. Estimated RMSE for the full grid and racing results.

While the racing approach selected the same candidate parameters as the complete grid for only 41.67% of the models, the performance metrics of the models selected by racing were nearly equal. The correlation of RMSE values was 0.968 and the rank correlation was 0.951. This indicates that, within a model, there were multiple tuning parameter combinations that had nearly identical results.

Finalizing a Model

Similar to what we have shown in previous chapters, the process of choosing the final model and fitting it on the training set is straightforward. The first step is to pick a workflow to finalize. Since the boosted tree model worked well, we'll extract that from the set, update the parameters with the numerically best settings, and fit to the training set:

```
best_results <-
  race_results %>%
  extract_workflow_set_result("boosting") %>%
  select_best(metric = "rmse")
best_results
#> # A tibble: 1 × 7
#>   trees min_n tree_depth learn_rate loss_reduction sample_size .config
#>   <int> <int>      <int>      <dbl>          <dbl>       <dbl> <chr>
#> 1  1957     8          7     0.0756    0.000000145       0.679 Preprocessor1_Model04
```

```
boosting_test_results <-
  race_results %>%
  extract_workflow("boosting") %>%
  finalize_workflow(best_results) %>%
  last_fit(split = concrete_split)
```

We can see the test set metrics results and visualize the predictions in Figure 15-5:

```
collect_metrics(boosting_test_results)
#> # A tibble: 2 × 4
#>   .metric .estimator .estimate .config
#>   <chr>   <chr>          <dbl> <chr>
#> 1 rmse    standard       3.33  Preprocessor1_Model1
#> 2 rsq     standard       0.956 Preprocessor1_Model1

boosting_test_results %>%
  collect_predictions() %>%
  ggplot(aes(x = compressive_strength, y = .pred)) +
  geom_abline(color = "gray50", lty = 2) +
  geom_point(alpha = 0.5) +
  coord_obs_pred() +
  labs(x = "observed", y = "predicted")
```

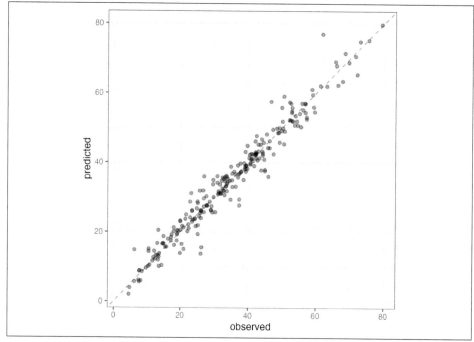

Figure 15-5. Observed versus predicted values for the test set.

We see here how well the observed and predicted compressive strength for these concrete mixtures align.

Chapter Summary

Often a data practitioner needs to consider a large number of possible modeling approaches for a task at hand, especially for new data sets and/or when there is little knowledge about what modeling strategy will work best. This chapter illustrated how to use workflow sets to investigate multiple models or feature engineering strategies in such a situation. Racing methods can more efficiently rank models than fitting every candidate model being considered.

## PART IV

Beyond the Basics

## CHAPTER 16

Dimensionality Reduction

Dimensionality reduction transforms a data set from a high-dimensional space into a low-dimensional space, and can be a good choice when you suspect there are "too many" variables. An excess of variables, usually predictors, can be a problem because it is difficult to understand or visualize data in higher dimensions.

What Problems Can Dimensionality Reduction Solve?

Dimensionality reduction can be used either in feature engineering or in exploratory data analysis. For example, in high-dimensional biology experiments, one of the first tasks, before any modeling, is to determine if there are any unwanted trends in the data (e.g., effects not related to the question of interest, such as lab-to-lab differences). Debugging the data is difficult when there are hundreds of thousands of dimensions, and dimensionality reduction can be an aid for exploratory data analysis.

Another potential consequence of having a multitude of predictors is possible harm to a model. The simplest example is a method like ordinary linear regression where the number of predictors should be less than the number of data points used to fit the model. Another issue is multicollinearity, where between-predictor correlations can negatively impact the mathematical operations used to estimate a model. If there are an extremely large number of predictors, it is fairly unlikely that there are an equal number of real underlying effects. Predictors may be measuring the same latent effect(s), and thus such predictors will be highly correlated. Many dimensionality reduction techniques thrive in this situation. In fact, most can be effective only when there are such relationships between predictors that can be exploited.

Principal component analysis (PCA) is one of the most straightforward methods for reducing the number of columns in the data set because it relies on linear methods and is unsupervised (i.e., does not consider the outcome data). For a high-dimensional classification problem, an initial plot of the main PCA components might show a clear separation between the classes. If this is the case, then it is fairly safe to assume that a linear classifier might do a good job. However, the converse is not true; a lack of separation does not mean that the problem is insurmountable.

When starting a new modeling project, reducing the dimensions of the data may provide some intuition about how hard the modeling problem may be.

The dimensionality reduction methods discussed in this chapter are generally *not* feature selection methods. Methods such as PCA represent the original predictors using a smaller subset of new features. All of the original predictors are required to compute these new features. The exception to this are sparse methods that have the ability to completely remove the impact of predictors when creating the new features.

This chapter has two goals:

- Demonstrate how to use recipes to create a small set of features that capture the main aspects of the original predictor set.
- Describe how recipes can be used on their own (as opposed to being used in a workflow object, as in Chapter 8).

The latter is helpful when testing or debugging a recipe. However, as described in Chapter 8, the best way to use a recipe for modeling is from within a workflow object.

In addition to the tidymodels package, this chapter uses the following packages: baguette, beans, bestNormalize, corrplot, discrim, embed, ggforce, klaR, learntidymodels (*https://oreil.ly/lyJtX*), mixOmics (*https://oreil.ly/DaXYl*), and uwot.

A Picture Is Worth a Thousand...Beans

Let's walk through how to use dimensionality reduction with recipes for an example data set. Koklu and Ozkan (2020) published a data set of visual characteristics of dried beans and described methods for determining the varieties of dried beans in an image. While the dimensionality of these data is not very large compared to many real-world modeling problems, it does provide a nice working example to demonstrate how to reduce the number of features. From their manuscript:

> The primary objective of this study is to provide a method for obtaining uniform seed varieties from crop production, which is in the form of population, so the seeds are not certified as a sole variety. Thus, a computer vision system was developed to distinguish

seven different registered varieties of dry beans with similar features in order to obtain uniform seed classification. For the classification model, images of 13,611 grains of 7 different registered dry beans were taken with a high-resolution camera.

Each image contains multiple beans. The process of determining which pixels correspond to a particular bean is called *image segmentation*. These pixels can be analyzed to produce features for each bean, such as color and morphology (i.e., shape). These features are then used to model the outcome (bean variety) because different bean varieties look different. The training data come from a set of manually labeled images, and this data set is used to create a predictive model that can distinguish between seven bean varieties: Cali, Horoz, Dermason, Seker, Bombay, Barbunya, and Sira. Producing an effective model can help manufacturers quantify the homogeneity of a batch of beans.

There are numerous methods for quantifying shapes of objects (Yang, Kpalma, and Ronsin 2008). Many are related to the boundaries or regions of the object of interest. Example of features include:

- The *area* (or size) can be estimated using the number of pixels in the object or the size of the convex hull around the object.

- We can measure the *perimeter* using the number of pixels in the boundary as well as the area of the bounding box (the smallest rectangle enclosing an object).

- The *major axis* quantifies the longest line connecting the most extreme parts of the object. The *minor axis* is perpendicular to the major axis.

- We can measure the *compactness* of an object using the ratio of the object's area to the area of a circle with the same perimeter. For example, the symbols "•" and "×" have very different compactness.

- There are also different measures of how *elongated* or oblong an object is. For example, the *eccentricity* statistic is the ratio of the major and minor axes. There are also related estimates for roundness and convexity.

Notice the eccentricity for the different shapes in Figure 16-1.

Shapes such as circles and squares have low eccentricity while oblong shapes have high values. Also, the metric is unaffected by the rotation of the object.

Many of these image features have high correlations; objects with large areas are more likely to have large perimeters. There are often multiple methods to quantify the same underlying characteristics (e.g., size).

In the bean data, 16 morphology features were computed: area, perimeter, major axis length, minor axis length, aspect ratio, eccentricity, convex area, equivalent diameter, extent, solidity, roundness, compactness, shape factor 1, shape factor 2, shape factor 3, and shape factor 4. The latter four are described in Symons and Fulcher (1988).

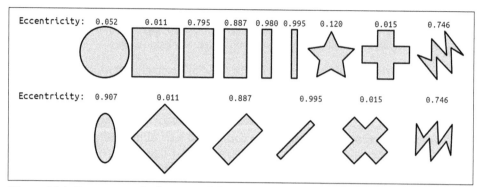

Figure 16-1. Some example shapes and their eccentricity statistics.

We can begin by loading the data:

```
library(tidymodels)
tidymodels_prefer()
library(beans)
```

 It is important to maintain good data discipline when evaluating dimensionality reduction techniques, especially if you will use them within a model.

For our analyses, we start by holding back a testing set with `initial_split()`. The remaining data are split into training and validation sets:

```
set.seed(1601)
bean_split <- initial_split(beans, strata = class, prop = 3/4)

bean_train <- training(bean_split)
bean_test  <- testing(bean_split)

set.seed(1602)
bean_val <- validation_split(bean_train, strata = class, prop = 4/5)
bean_val$splits[[1]]
#> <Training/Validation/Total>
#> <8163/2043/10206>
```

To visually assess how well different methods perform, we can estimate the methods on the training set (n = 8,163 beans) and display the results using the validation set (n = 2,043).

Before beginning any dimensionality reduction, we can spend some time investigating our data. Since we know that many of these shape features are probably measuring similar concepts, let's take a look at the correlation structure of the data in Figure 16-2 using this code:

```
library(corrplot)
tmwr_cols <- colorRampPalette(c("#91CBD765", "#CA225E"))
bean_train %>%
  select(-class) %>%
  cor() %>%
  corrplot(col = tmwr_cols(200), tl.col = "black", method = "ellipse")
```

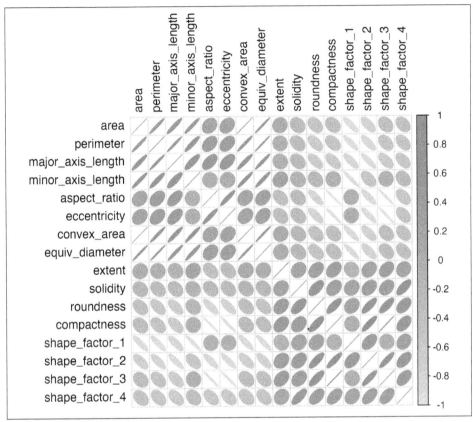

Figure 16-2. Correlation matrix of the predictors with variables ordered via clustering.

Many of these predictors are highly correlated, such as area and perimeter or shape factors 2 and 3. While we don't take the time to do it here, it is also important to see if this correlation structure significantly changes across the outcome categories. This can help create better models.

A Starter Recipe

It's time to look at the beans data in a smaller space. We can start with a basic recipe to preprocess the data prior to any dimensionality reduction steps. Several predictors are ratios and so are likely to have skewed distributions. Such distributions can wreak havoc on variance calculations (such as the ones used in PCA). The bestNormalize package (*https://oreil.ly/v26pw*) has a step that can enforce a symmetric distribution for the predictors. We'll use this to mitigate the issue of skewed distributions:

```
library(bestNormalize)
bean_rec <-
  # Use the training data from the bean_val split object
  recipe(class ~ ., data = analysis(bean_val$splits[[1]])) %>%
  step_zv(all_numeric_predictors()) %>%
  step_orderNorm(all_numeric_predictors()) %>%
  step_normalize(all_numeric_predictors())
```

 Remember that when invoking the recipe() function, the steps are not estimated or executed in any way.

This recipe will be extended with additional steps for the dimensionality reduction analyses. Before that, let's go over how a recipe can be used outside of a workflow.

Recipes in the Wild

As mentioned in Chapter 8, a workflow containing a recipe uses fit() to estimate the recipe and model, then predict() to process the data and make model predictions. There are analogous functions in the recipes package that can be used for the same purpose:

- prep(recipe, training) fits the recipe to the training set.
- bake(recipe, new_data) applies the recipe operations to new_data.

Figure 16-3 summarizes this. Let's look at each of these functions in more detail.

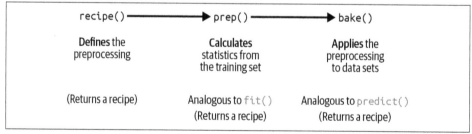

Figure 16-3. Summary of recipe-related functions.

Preparing a Recipe

Let's estimate `bean_rec` using the training set data, with `prep(bean_rec)`:

```
bean_rec_trained <- prep(bean_rec)
bean_rec_trained
#> Recipe
#>
#> Inputs:
#>
#>        role #variables
#>     outcome           1
#>   predictor          16
#>
#> Training data contained 8163 data points and no missing data.
#>
#> Operations:
#>
#> Zero variance filter removed <none> [trained]
#> orderNorm transformation on area, perimeter, major_axis_length, minor_axis... [trained]
#> Centering and scaling for area, perimeter, major_axis_length, minor_axis_... [trained]
```

 Remember that `prep()` for a recipe is like `fit()` for a model.

Note in the output that the steps have been trained and that the selectors are no longer general (i.e., `all_numeric_predictors()`); they now show the actual columns that were selected. Also, `prep(bean_rec)` does not require the `training` argument. You can pass any data into that argument, but omitting it means that the original data from the call to `recipe()` will be used. In our case, this was the training set data.

One important argument to `prep()` is `retain`. When `retain = TRUE` (the default), the estimated version of the training set is kept within the recipe. This data set has been preprocessed using all of the steps listed in the recipe. Since `prep()` has to execute the recipe as it proceeds, it may be advantageous to keep this version of the training set so that, if that data set is to be used later, redundant calculations can be avoided. However, if the training set is big, it may be problematic to keep such a large amount of data in memory. Use `retain = FALSE` to avoid this.

Once new steps are added to this estimated recipe, reapplying `prep()` will estimate only the untrained steps. This will come in handy when we try different feature extraction methods.

If you encounter errors when working with a recipe, prep() can be used with its verbose option to troubleshoot:

```
bean_rec_trained %>%
  step_dummy(cornbread) %>%  # <- not a real predictor
  prep(verbose = TRUE)
#> oper 1 step zv [pre-trained]
#> oper 2 step orderNorm [pre-trained]
#> oper 3 step normalize [pre-trained]
#> oper 4 step dummy [training]
#> Error in `chr_as_locations()`:
#> ! Can't subset columns that don't exist.
#> ✖ Column `cornbread` doesn't exist.
```

Another option that can help you understand what happens in the analysis is log_changes:

```
show_variables <-
  bean_rec %>%
  prep(log_changes = TRUE)
#> step_zv (zv_6JtxV): same number of columns
#>
#> step_orderNorm (orderNorm_4r8al): same number of columns
#>
#> step_normalize (normalize_x6oqH): same number of columns
```

Baking the Recipe

Using bake() with a recipe is much like using predict() with a model; the operations estimated from the training set are applied to any data, like testing data or new data at prediction time.

For example, the validation set samples can be processed:

```
bean_validation <- bean_val$splits %>% pluck(1) %>% assessment()
bean_val_processed <- bake(bean_rec_trained, new_data = bean_validation)
```

Figure 16-4 shows histograms of the area predictor before and after the recipe was prepared:

```
library(patchwork)
p1 <-
  bean_validation %>%
  ggplot(aes(x = area)) +
  geom_histogram(bins = 30, color = "white", fill = "blue", alpha = 1/3) +
  ggtitle("Original validation set data")

p2 <-
  bean_val_processed %>%
  ggplot(aes(x = area)) +
```

```
geom_histogram(bins = 30, color = "white", fill = "red", alpha = 1/3) +
ggtitle("Processed validation set data")
```

```
p1 + p2
```

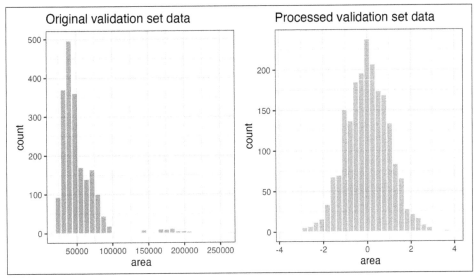

Figure 16-4. The area predictor before and after preprocessing.

Two important aspects of bake() are worth noting here.

First, as previously mentioned, using prep(recipe, retain = TRUE) keeps the existing processed version of the training set in the recipe. This enables the user to use bake(recipe, new_data = NULL), which returns that data set without further computations. For example:

```
bake(bean_rec_trained, new_data = NULL) %>% nrow()
#> [1] 8163
bean_val$splits %>% pluck(1) %>% analysis() %>% nrow()
#> [1] 8163
```

If the training set is not pathologically large, using this value of retain can save a lot of computational time.

Second, additional selectors can be used in the call to specify which columns to return. The default selector is everything(), but more specific directives can be used.

We will use prep() and bake() in the next section to illustrate some of these options.

Feature Extraction Techniques

Since recipes are the primary option in tidymodels for dimensionality reduction, let's write a function that will estimate the transformation and plot the resulting data:

```
plot_validation_results <- function(recipe, dat = assessment(bean_val$splits[[1]])) {
  set.seed(1)
  plot_data <-
    recipe %>%
    # Estimate any additional steps
    prep() %>%
    # Process the data (the validation set by default)
    bake(new_data = dat, all_predictors(), all_outcomes()) %>%
    # Sample the data down to be more readable
    sample_n(250)

  # Convert feature names to symbols to use with quasiquotation
  nms <- names(plot_data)
  x_name <- sym(nms[1])
  y_name <- sym(nms[2])

  plot_data %>%
    ggplot(aes(x = !!x_name, y = !!y_name, col = class,
               fill = class, pch = class)) +
    geom_point(alpha = 0.9) +
    scale_shape_manual(values = 1:7) +
    # Make equally sized axes
    coord_obs_pred() +
    theme_bw()
}
```

We will reuse this function several times in this chapter.

A series of several feature extraction methodologies are explored here. An overview of most can be found in section 6.3.1 of Kuhn and Johnson (2020) (*https://oreil.ly/xllmg*)and the references therein. The UMAP method is described in McInnes, Healy, and Melville (2020).

Principal Component Analysis

We've mentioned PCA several times already in this book, and it's time to go into more detail. PCA is an unsupervised method that uses linear combinations of the predictors to define new features. These features attempt to account for as much variation as possible in the original data. We add `step_pca()` to the original recipe and use our function to visualize the results on the validation set in Figure 16-5 using:

```
bean_rec_trained %>%
  step_pca(all_numeric_predictors(), num_comp = 4) %>%
  plot_validation_results() +
  ggtitle("Principal Component Analysis")
```

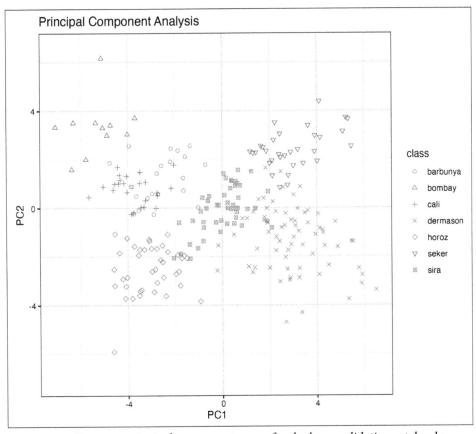

Figure 16-5. First two principal component scores for the bean validation set, by class.

We see that the first two components PC1 and PC2, especially when used together, do an effective job distinguishing between or separating the classes. This may lead us to expect that the overall problem of classifying these beans will not be especially difficult.

Recall that PCA is unsupervised. For these data, it turns out that the PCA components that explain the most variation in the predictors also happen to be predictive of the classes. What features are driving performance? The learntidymodels package has functions that can help visualize the top features for each component. We'll need the prepared recipe; the PCA step is added in the following code along with a call to prep():

```
library(learntidymodels)
bean_rec_trained %>%
  step_pca(all_numeric_predictors(), num_comp = 4) %>%
  prep() %>%
  plot_top_loadings(component_number <= 4, n = 5) +
```

```
scale_fill_brewer(palette = "Paired") +
ggtitle("Principal Component Analysis")
```

This produces Figure 16-6.

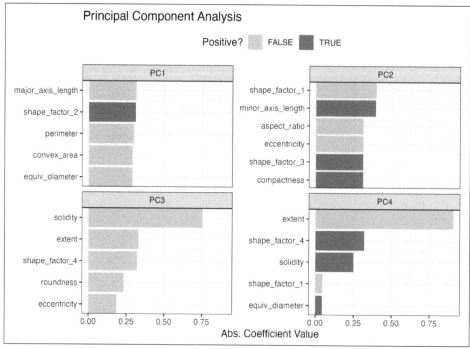

Figure 16-6. Predictor loadings for the PCA transformation.

The top loadings are mostly related to the cluster of correlated predictors shown in the top-left portion of the previous correlation plot: perimeter, area, major axis length, and convex area. These are all related to bean size. Shape factor 2, from Symons and Fulcher (1988), is the area over the cube of the major axis length and is therefore also related to bean size. Measures of elongation appear to dominate the second PCA component.

Partial Least Squares

PLS, which we introduced in "Submodel Optimization" on page 199, is a supervised version of PCA. It tries to find components that simultaneously maximize the variation in the predictors while also maximizing the relationship between those components and the outcome. Figure 16-7 shows the results of this slightly modified version of the PCA code:

```
bean_rec_trained %>%
  step_pls(all_numeric_predictors(), outcome = "class", num_comp = 4) %>%
  plot_validation_results() +
  ggtitle("Partial Least Squares")
```

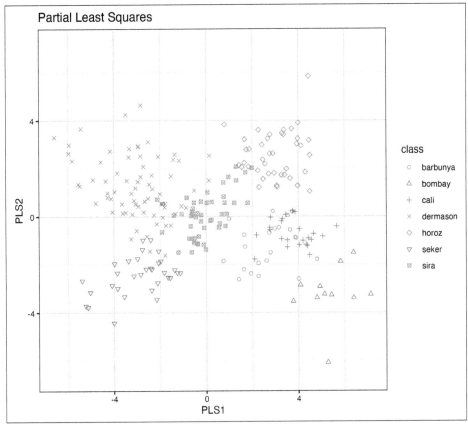

Figure 16-7. First two PLS component scores for the bean validation set, by class.

The first two PLS components plotted in Figure 16-7 are nearly identical to the first two PCA components! We find this result because those PCA components are so effective at separating the varieties of beans. The remaining components are different. Figure 16-8 visualizes the loadings, the top features for each component:

```
bean_rec_trained %>%
  step_pls(all_numeric_predictors(), outcome = "class", num_comp = 4) %>%
  prep() %>%
  plot_top_loadings(component_number <= 4, n = 5, type = "pls") +
  scale_fill_brewer(palette = "Paired") +
  ggtitle("Partial Least Squares")
```

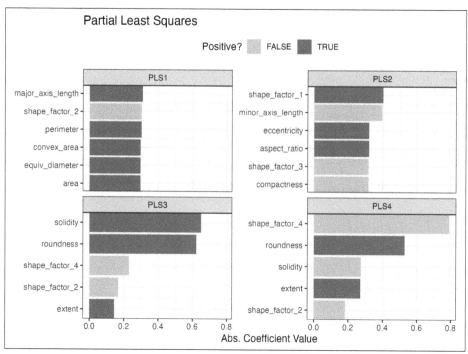

Figure 16-8. Predictor loadings for the PLS transformation.

Solidity (i.e., the density of the bean) drives the third PLS component, along with roundness. Solidity may be capturing bean features related to "bumpiness" of the bean surface since it can measure irregularity of the bean boundaries.

Independent Component Analysis

ICA is slightly different than PCA in that it finds components that are as statistically independent from one another as possible (as opposed to being uncorrelated). It can be thought of as maximizing the "non-Gaussianity" of the ICA components, or separating information instead of compressing information like PCA. Let's use `step_ica()` to produce Figure 16-9:

```
bean_rec_trained %>%
  step_ica(all_numeric_predictors(), num_comp = 4) %>%
  plot_validation_results() +
  ggtitle("Independent Component Analysis")
```

Inspecting this plot, there does not appear to be much separation between the classes in the first few components when using ICA. These independent (or as independent as possible) components do not separate the bean types.

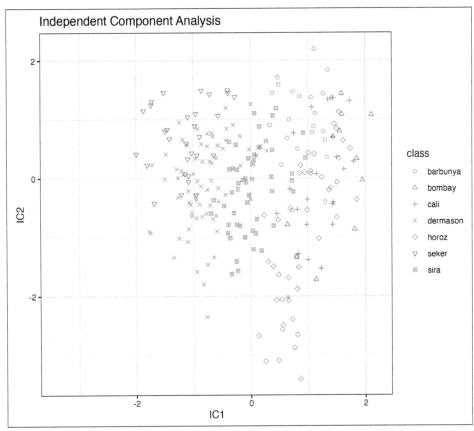

class

- ○ barbunya
- △ bombay
- + cali
- × dermason
- ◇ horoz
- ▽ seker
- ⊠ sira

Figure 16-9. First two ICA component scores for the bean validation set, by class.

Uniform Manifold Approximation and Projection

UMAP is similar to the popular t-SNE method for nonlinear dimension reduction. In the original high-dimensional space, UMAP uses a distance-based nearest neighbor method to find local areas of the data where the data points are more likely to be related. The relationship between data points is saved as a directed graph model where most points are not connected.

From there, UMAP translates points in the graph to the reduced dimensional space. To do this, the algorithm has an optimization process that uses cross-entropy to map data points to the smaller set of features so that the graph is well approximated.

To create the mapping, the embed package contains a step function for this method, visualized in Figure 16-10:

```
library(embed)
bean_rec_trained %>%
  step_umap(all_numeric_predictors(), num_comp = 4) %>%
```

```
plot_validation_results() +
  ggtitle("UMAP")
```

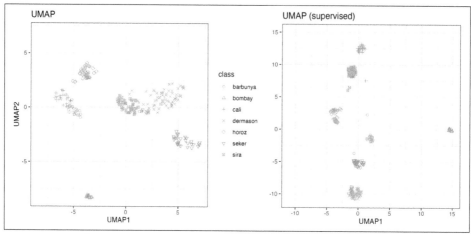

Figure 16-10. The first two UMAP component scores for the bean validation set, by class. Results are shown for unsupervised and supervised versions.

The resulting plot is shown on the left-hand side of Figure 16-10. While the between-cluster space is pronounced, the clusters can contain a heterogeneous mixture of classes.

There is also a supervised version of UMAP:

```
bean_rec_trained %>%
  step_umap(all_numeric_predictors(), outcome = "class", num_comp = 4) %>%
  plot_validation_results() +
  ggtitle("UMAP (supervised)")
```

The supervised method shown in Figure 16-10 looks promising for modeling the data.

UMAP is a powerful method to reduce the feature space. However, it can be very sensitive to tuning parameters (e.g., the number of neighbors and so on). For this reason, it would help to experiment with a few of the parameters to assess how robust the results are for these data.

Modeling

Both the PLS and UMAP methods are worth investigating in conjunction with different models. Let's explore a variety of different models with these dimensionality reduction techniques (along with no transformation at all): a single-layer neural network, bagged trees, flexible discriminant analysis (FDA), naive Bayes, and regularized discriminant analysis (RDA).

Now that we are back in "modeling mode," we'll create a series of model specifications and then use a workflow set to tune the models in the following code. Note that the model parameters are tuned in conjunction with the recipe parameters (e.g., size of the reduced dimension, UMAP parameters):

```
library(baguette)
library(discrim)

mlp_spec <-
  mlp(hidden_units = tune(), penalty = tune(), epochs = tune()) %>%
  set_engine('nnet') %>%
  set_mode('classification')

bagging_spec <-
  bag_tree() %>%
  set_engine('rpart') %>%
  set_mode('classification')

fda_spec <-
  discrim_flexible(
    prod_degree = tune()
  ) %>%
  set_engine('earth')

rda_spec <-
  discrim_regularized(frac_common_cov = tune(), frac_identity = tune()) %>%
  set_engine('klaR')

bayes_spec <-
  naive_Bayes() %>%
  set_engine('klaR')
```

We also need recipes for the dimensionality reduction methods we'll try. Let's start with a base recipe bean_rec and then extend it with different dimensionality reduction steps:

```
bean_rec <-
  recipe(class ~ ., data = bean_train) %>%
  step_zv(all_numeric_predictors()) %>%
  step_orderNorm(all_numeric_predictors()) %>%
  step_normalize(all_numeric_predictors())

pls_rec <-
  bean_rec %>%
  step_pls(all_numeric_predictors(), outcome = "class", num_comp = tune())

umap_rec <-
  bean_rec %>%
  step_umap(
    all_numeric_predictors(),
    outcome = "class",
    num_comp = tune(),
    neighbors = tune(),
    min_dist = tune()
  )
```

Once again, the workflowsets package takes the preprocessors and models and crosses them. The `control` option `parallel_over` is set so that the parallel processing can work simultaneously across tuning parameter combinations. The `workflow_map()` function applies grid search to optimize the model/preprocessing parameters (if any) across 10 parameter combinations. The multiclass area under the ROC curve is estimated on the validation set:

```
ctrl <- control_grid(parallel_over = "everything")
bean_res <-
  workflow_set(
    preproc = list(basic = class ~., pls = pls_rec, umap = umap_rec),
    models = list(bayes = bayes_spec, fda = fda_spec,
                  rda = rda_spec, bag = bagging_spec,
                  mlp = mlp_spec)
  ) %>%
  workflow_map(
    verbose = TRUE,
    seed = 1603,
    resamples = bean_val,
    grid = 10,
    metrics = metric_set(roc_auc),
    control = ctrl
  )
```

We can rank the models by their validation set estimates of the area under the ROC curve:

```
rankings <-
  rank_results(bean_res, select_best = TRUE) %>%
  mutate(method = map_chr(wflow_id, ~ str_split(.x, "_", simplify = TRUE)[1]))

tidymodels_prefer()
filter(rankings, rank <= 5) %>% dplyr::select(rank, mean, model, method)
#> # A tibble: 5 × 4
#>    rank  mean model               method
#>   <int> <dbl> <chr>               <chr>
#> 1     1 0.995 mlp                 basic
#> 2     2 0.995 discrim_regularized pls
#> 3     3 0.994 mlp                 pls
#> 4     4 0.994 naive_Bayes         pls
#> 5     5 0.994 discrim_flexible    basic
```

Figure 16-11 illustrates this ranking.

It is clear from these results that most models give very good performance; there are few bad choices here. For demonstration, we'll use the RDA model with PLS features as the final model. We will finalize the workflow with the numerically best parameters, fit it to the training set, then evaluate with the test set:

```
rda_res <-
  bean_res %>%
  extract_workflow("pls_rda") %>%
  finalize_workflow(
    bean_res %>%
      extract_workflow_set_result("pls_rda") %>%
      select_best(metric = "roc_auc")
```

```
    ) %>%
    last_fit(split = bean_split, metrics = metric_set(roc_auc))

rda_wflow_fit <- rda_res$.workflow[[1]]
```

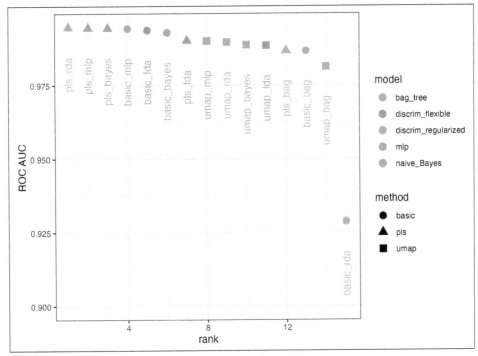

Figure 16-11. Area under the ROC curve from the validation set.

What are the results for our metric (multiclass ROC AUC) on the testing set?

```
collect_metrics(rda_res)
#> # A tibble: 1 × 4
#>    .metric .estimator .estimate .config
#>    <chr>   <chr>          <dbl> <chr>
#> 1 roc_auc hand_till       0.995 Preprocessor1_Model1
```

Pretty good! We'll use this model in Chapter 18 to demonstrate variable importance methods.

Chapter Summary

Dimensionality reduction can be a helpful method for exploratory data analysis as well as modeling. The recipes and embed packages contain steps for a variety of different methods and workflowsets, which facilitates choosing an appropriate method for a data set. This chapter also discussed how recipes can be used on their own, either for debugging problems with a recipe or directly for exploratory data analysis and data visualization.

## CHAPTER 17

Encoding Categorical Data

For statistical modeling in R, the preferred representation for categorical or nominal data is a *factor*, a variable that can take on a limited number of different values; internally, factors are stored as a vector of integer values together with a set of text labels.[1] In Chapter 8 we introduced feature engineering approaches, including those to encode or transform qualitative or nominal data into a representation better suited for most model algorithms. We discussed how to transform a categorical variable, such as the Bldg_Type in our Ames housing data (with levels OneFam, TwoFmCon, Duplex, Twnhs, and TwnhsE), to a set of dummy or indicator variables like those shown in Table 17-1.

Table 17-1. Illustration of binary encodings (i.e., dummy variables) for a qualitative predictor

Raw data	TwoFmCon	Duplex	Twnhs	TwnhsE
OneFam	0	0	0	0
TwoFmCon	1	0	0	0
Duplex	0	1	0	0
Twnhs	0	0	1	0
TwnhsE	0	0	0	1

Many model implementations require such a transformation to a numeric representation for categorical data.

---

1 This is in contrast to statistical modeling in Python, where categorical variables are often directly represented by integers alone, such as 0, 1, 2 representing red, blue, and green.

The Appendix presents a table of recommended preprocessing techniques for different models; notice how many of the models in the table require a numeric encoding for all predictors.

However, for some realistic data sets, straightforward dummy variables are not a good fit. This often happens because there are *too many* categories or there are *new* categories at prediction time. In this chapter, we discuss more sophisticated options for encoding categorical predictors that address these issues. These options are available as tidymodels recipe steps in the embed (*https://oreil.ly/lQfup*) and textrecipes (*https://oreil.ly/PHwTv*) packages.

Is an Encoding Necessary?

A minority of models, such as those based on trees or rules, can handle categorical data natively and do not require encoding or transformation of these kinds of features. A tree-based model can natively partition a variable like `Bldg_Type` into groups of factor levels, perhaps `OneFam` alone in one group and `Duplex` and `Twnhs` together in another group. Naive Bayes models are another example where the structure of the model can deal with categorical variables natively; distributions are computed within each level, for example, for all the different kinds of `Bldg_Type` in the data set.

Models that can handle categorical features natively can *also* deal with numeric, continuous features, making the transformation or encoding of such variables optional. Does this help in some way, perhaps with model performance or time to train models? Typically no, as section 5.7 of Kuhn and Johnson (2020) (*https://oreil.ly/0ImIU*) shows, using benchmark data sets with untransformed factor variables compared with transformed dummy variables for those same features. Using dummy encodings did not typically result in better model performance but often required more time to train the models.

We advise starting with untransformed categorical variables when a model allows it; note that more complex encodings often do not result in better performance for such models.

Encoding Ordinal Predictors

Sometimes qualitative columns can be *ordered*, such as "low," "medium," and "high." In base R, the default encoding strategy is to make new numeric columns that are polynomial expansions of the data. For columns that have five ordinal values, like the example shown in Table 17-2, the factor column is replaced with columns for linear, quadratic, cubic, and quartic terms.

Table 17-2. Polynominal expansions for encoding an ordered variable.

Raw data	Linear	Quadratic	Cubic	Quartic
None	−0.63	0.53	−0.32	0.12
A little	−0.32	−0.27	0.63	−0.48
Some	0.00	−0.53	0.00	0.72
A bunch	0.32	−0.27	−0.63	−0.48
Copious amounts	0.63	0.53	0.32	0.12

While this is not unreasonable, it is not an approach that people tend to find useful. For example, an 11-degree polynomial is probably not the most effective way of encoding an ordinal factor for the months of the year. Instead, consider trying recipe steps related to ordered factors, such as `step_unorder()`, to convert to regular factors, and `step_ordinalscore()`, which maps specific numeric values to each factor level.

Using the Outcome for Encoding Predictors

There are multiple options for encodings more complex than dummy or indicator variables. One method called *effect* or *likelihood encodings* replaces the original categorical variables with a single numeric column that measures the effect of those data (Micci-Barreca 2001; Zumel and Mount 2019). For example, for the neighborhood predictor in the Ames housing data, we can compute the mean or median sale price for each neighborhood (as shown in Figure 17-1) and substitute these means for the original data values:

```
ames_train %>%
  group_by(Neighborhood) %>%
  summarize(mean = mean(Sale_Price),
            std_err = sd(Sale_Price) / sqrt(length(Sale_Price))) %>%
  ggplot(aes(y = reorder(Neighborhood, mean), x = mean)) +
  geom_point() +
  geom_errorbar(aes(xmin = mean - 1.64 * std_err, xmax = mean + 1.64 * std_err)) +
  labs(y = NULL, x = "Price (mean, log scale)")
```

This kind of effect encoding works well when your categorical variable has many levels.

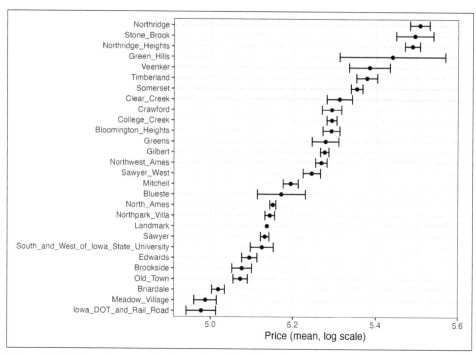

Figure 17-1. Mean home price for neighborhoods in the Ames training set, which can be used as an effect encoding for this categorical variable.

Effect Encodings in tidymodels

In tidymodels, the embed package includes several recipe step functions for different kinds of effect encodings, such as `step_lencode_glm()`, `step_lencode_mixed()`, and `step_lencode_bayes()`. These steps use a generalized linear model to estimate the effect of each level in a categorical predictor on the outcome. When using a recipe step like `step_lencode_glm()`, specify the variable being encoded first and then the outcome using `vars()`:

```
library(embed)

ames_glm <-
  recipe(Sale_Price ~ Neighborhood + Gr_Liv_Area + Year_Built + Bldg_Type +
           Latitude + Longitude, data = ames_train) %>%
  step_log(Gr_Liv_Area, base = 10) %>%
  step_lencode_glm(Neighborhood, outcome = vars(Sale_Price)) %>%
  step_dummy(all_nominal_predictors()) %>%
  step_interact( ~ Gr_Liv_Area:starts_with("Bldg_Type_") ) %>%
  step_ns(Latitude, Longitude, deg_free = 20)

ames_glm
#> Recipe
#>
#> Inputs:
```

```
#>
#>        role #variables
#>     outcome           1
#>  predictor            6
#>
#> Operations:
#>
#> Log transformation on Gr_Liv_Area
#> Linear embedding for factors via GLM for Neighborhood
#> Dummy variables from all_nominal_predictors()
#> Interactions with Gr_Liv_Area:starts_with("Bldg_Type_")
#> Natural splines on Latitude, Longitude
```

As detailed in Chapter 16, we can `prep()` our recipe to fit or estimate parameters for the preprocessing transformations using training data. We can then `tidy()` this prepared recipe to see the results:

```
glm_estimates <-
  prep(ames_glm) %>%
  tidy(number = 2)

glm_estimates
#> # A tibble: 29 × 4
#>    level                value terms        id
#>    <chr>                <dbl> <chr>        <chr>
#> 1 North_Ames            5.15 Neighborhood lencode_glm_ZsXdy
#> 2 College_Creek         5.29 Neighborhood lencode_glm_ZsXdy
#> 3 Old_Town              5.07 Neighborhood lencode_glm_ZsXdy
#> 4 Edwards               5.09 Neighborhood lencode_glm_ZsXdy
#> 5 Somerset              5.35 Neighborhood lencode_glm_ZsXdy
#> 6 Northridge_Heights    5.49 Neighborhood lencode_glm_ZsXdy
#> # … with 23 more rows
```

When we use the newly encoded `Neighborhood` numeric variable created via this method, we substitute the original level (such as `"North_Ames"`) with the estimate for `Sale_Price` from the GLM.

Effect encoding methods like this one can also seamlessly handle situations where a novel factor level is encountered in the data. This `value` is the predicted price from the GLM when we don't have any specific neighborhood information:

```
glm_estimates %>%
  filter(level == "..new")
#> # A tibble: 1 × 4
#>   level value terms        id
#>   <chr> <dbl> <chr>        <chr>
#> 1 ..new  5.23 Neighborhood lencode_glm_ZsXdy
```

 Effect encodings can be powerful but should be used with care. The effects should be computed from the training set, after data splitting. This type of supervised preprocessing should be rigorously resampled to avoid overfitting (see Chapter 10).

When you create an effect encoding for your categorical variable, you are effectively layering a mini-model inside your actual model. The possibility of overfitting with effect encodings is a representative example for why feature engineering *must* be considered part of the model process, as described in Chapter 7, and why feature engineering must be estimated together with model parameters inside resampling.

Effect Encodings with Partial Pooling

Creating an effect encoding with `step_lencode_glm()` estimates the effect separately for each factor level (in this example, neighborhood). However, some of these neighborhoods have many houses in them, and some have only a few. There is much more uncertainty in our measurement of price for the single training set home in the Landmark neighborhood than the 354 training set homes in North Ames. We can use *partial pooling* to adjust these estimates so that levels with small sample sizes are shrunken toward the overall mean. The effects for each level are modeled all at once using a mixed or hierarchical generalized linear model:

```
ames_mixed <-
  recipe(Sale_Price ~ Neighborhood + Gr_Liv_Area + Year_Built + Bldg_Type +
           Latitude + Longitude, data = ames_train) %>%
  step_log(Gr_Liv_Area, base = 10) %>%
  step_lencode_mixed(Neighborhood, outcome = vars(Sale_Price)) %>%
  step_dummy(all_nominal_predictors()) %>%
  step_interact( ~ Gr_Liv_Area:starts_with("Bldg_Type_") ) %>%
  step_ns(Latitude, Longitude, deg_free = 20)

ames_mixed
#> Recipe
#>
#> Inputs:
#>
#>        role #variables
#>     outcome          1
#>   predictor          6
#>
#> Operations:
#>
#> Log transformation on Gr_Liv_Area
#> Linear embedding for factors via mixed effects for Neighborhood
#> Dummy variables from all_nominal_predictors()
#> Interactions with Gr_Liv_Area:starts_with("Bldg_Type_")
#> Natural splines on Latitude, Longitude
```

Let's `prep()` and `tidy()` this recipe to see the results:

```
mixed_estimates <-
  prep(ames_mixed) %>%
  tidy(number = 2)

mixed_estimates
#> # A tibble: 29 × 4
#>    level           value terms        id
#>    <chr>           <dbl> <chr>        <chr>
#> 1 North_Ames       5.15 Neighborhood lencode_mixed_SC9hi
```

```
#> 2 College_Creek         5.29 Neighborhood lencode_mixed_SC9hi
#> 3 Old_Town              5.07 Neighborhood lencode_mixed_SC9hi
#> 4 Edwards               5.10 Neighborhood lencode_mixed_SC9hi
#> 5 Somerset              5.35 Neighborhood lencode_mixed_SC9hi
#> 6 Northridge_Heights    5.49 Neighborhood lencode_mixed_SC9hi
#> # … with 23 more rows
```

New levels are then encoded at close to the same value as with the GLM:

```
mixed_estimates %>%
  filter(level == "..new")
#> # A tibble: 1 × 4
#>   level value terms        id
#>   <chr> <dbl> <chr>        <chr>
#> 1 ..new  5.23 Neighborhood lencode_mixed_SC9hi
```

 You can use a fully Bayesian hierarchical model for the effects in the same way with step_lencode_bayes().

Let's visually compare the effects using partial pooling versus no pooling in Figure 17-2:

```
glm_estimates %>%
  rename(`no pooling` = value) %>%
  left_join(
    mixed_estimates %>%
      rename(`partial pooling` = value), by = "level"
  ) %>%
  left_join(
    ames_train %>%
      count(Neighborhood) %>%
      mutate(level = as.character(Neighborhood))
  ) %>%
  ggplot(aes(`no pooling`, `partial pooling`, size = sqrt(n))) +
  geom_abline(color = "gray50", lty = 2) +
  geom_point(alpha = 0.7) +
  coord_fixed()
#> Warning: Removed 1 rows containing missing values (geom_point).
```

Notice in Figure 17-2 that most estimates for neighborhood effects are about the same when we compare pooling to no pooling. However, the neighborhoods with the fewest homes in them have been pulled (either up or down) toward the mean effect. When we use pooling, we shrink the effect estimates toward the mean because we don't have as much evidence about the price in those neighborhoods.

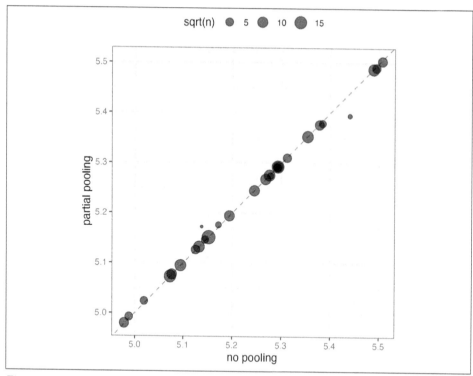

Figure 17-2. Comparing the effect encodings for neighborhoods estimated without pooling to those with partial pooling.

Feature Hashing

Traditional dummy variables as described in Chapter 8 require that all of the possible categories be known to create a full set of numeric features. *Feature hashing* methods (Weinberger et al. 2009) also create dummy variables but only consider the value of the category to assign it to a predefined pool of dummy variables. Let's look at the Neighborhood values in Ames again and use the rlang::hash() function to understand more:

```
library(rlang)

ames_hashed <-
  ames_train %>%
  mutate(Hash = map_chr(Neighborhood, hash))

ames_hashed %>%
  select(Neighborhood, Hash)
#> # A tibble: 2,342 × 2
#>    Neighborhood   Hash
#>    <fct>          <chr>
#> 1 North_Ames     076543f71313e522efe157944169d919
#> 2 North_Ames     076543f71313e522efe157944169d919
```

```
#> 3 Briardale       b598bec306983e3e68a3118952df8cf0
#> 4 Briardale       b598bec306983e3e68a3118952df8cf0
#> 5 Northpark_Villa 6af95b5db968bf393e78188a81e0e1e4
#> 6 Northpark_Villa 6af95b5db968bf393e78188a81e0e1e4
#> # … with 2,336 more rows
```

If we input Briardale to this hashing function, we will always get the same output. The neighborhoods in this case are called the *keys*, while the outputs are the *hashes*.

A hashing function takes an input of variable size and maps it to an output of fixed size. Hashing functions are commonly used in cryptography and databases.

The `rlang::hash()` function generates a 128-bit hash, which means there are 2^128 possible hash values. This is great for some applications but doesn't help with feature hashing of *high-cardinality* variables (variables with many levels). In feature hashing, the number of possible hashes is a hyperparameter and is set by the model developer through computing the modulo of the integer hashes. We can get 16 possible hash values by using `Hash %% 16`:

```
ames_hashed %>%
    ## first make a smaller hash for integers that R can handle
    mutate(Hash = strtoi(substr(Hash, 26, 32), base = 16L),
           ## now take the modulo
           Hash = Hash %% 16) %>%
    select(Neighborhood, Hash)
#> # A tibble: 2,342 × 2
#>   Neighborhood    Hash
#>   <fct>           <dbl>
#> 1 North_Ames         9
#> 2 North_Ames         9
#> 3 Briardale          0
#> 4 Briardale          0
#> 5 Northpark_Villa    4
#> 6 Northpark_Villa    4
#> # … with 2,336 more rows
```

Now instead of the 28 neighborhoods in our original data or an incredibly huge number of the original hashes, we have 16 hash values. This method is very fast and memory efficient, and it can be a good strategy when there are a large number of possible categories.

Feature hashing is useful for text data as well as high-cardinality categorical data. See section 6.7 of Hvitfeldt and Silge (2021) (*https://oreil.ly/mN7fo*) for a case study demonstration with text predictors.

We can implement feature hashing using a tidymodels recipe step from the textre-cipes package:

```
library(textrecipes)
ames_hash <-
  recipe(Sale_Price ~ Neighborhood + Gr_Liv_Area + Year_Built + Bldg_Type +
            Latitude + Longitude, data = ames_train) %>%
  step_log(Gr_Liv_Area, base = 10) %>%
  step_dummy_hash(Neighborhood, signed = FALSE, num_terms = 16L) %>%
  step_dummy(all_nominal_predictors()) %>%
  step_interact( ~ Gr_Liv_Area:starts_with("Bldg_Type_") ) %>%
  step_ns(Latitude, Longitude, deg_free = 20)

ames_hash
#> Recipe
#>
#> Inputs:
#>
#>        role #variables
#>     outcome          1
#>   predictor          6
#>
#> Operations:
#>
#> Log transformation on Gr_Liv_Area
#> Feature hashing with Neighborhood
#> Dummy variables from all_nominal_predictors()
#> Interactions with Gr_Liv_Area:starts_with("Bldg_Type_")
#> Natural splines on Latitude, Longitude
```

Feature hashing is fast and efficient but has a few downsides. For example, different category values often map to the same hash value. This is called a *collision* or *aliasing*. How often did this happen with our neighborhoods in Ames? Table 17-3 presents the distribution of number of neighborhoods per hash value.

Table 17-3. The number of hash features at each number of neighborhoods

Number of neighborhoods within a hash feature	Number of occurrences
0	1
1	7
2	4
4	1

The number of neighborhoods mapped to each hash value varies between zero and four. All of the hash values greater than one are examples of hash collisions.

What are some things to consider when using feature hashing?

- Feature hashing is not directly interpretable because hash functions cannot be reversed. We can't determine what the input category levels were from the hash value, or if a collision occurred.

- The number of hash values is a *tuning parameter* of this preprocessing technique, and you should try several values to determine what is best for your particular modeling approach. A lower number of hash values results in more collisions, but a high number may not be an improvement over your original high-cardinality variable.
- Feature hashing can handle new category levels at prediction time, since it does not rely on predetermined dummy variables.
- You can reduce hash collisions with a *signed* hash by using `signed = TRUE`. This expands the values from only 1 to either +1 or –1, depending on the sign of the hash.

> It is likely that some hash columns will contain all zeros, as we see in this example. We recommend a zero-variance filter via `step_zv()` to filter out such columns.

More Encoding Options

Even more options are available for transforming factors to a numeric representation.

We can build a full set of *entity embeddings* (Guo and Berkhahn 2016) to transform a categorical variable with many levels to a set of lower-dimensional vectors. This approach is best suited to a nominal variable with many category levels, many more than the example we've used with neighborhoods in Ames.

> The idea of entity embeddings comes from the methods used to create word embeddings from text data. See Chapter 5 of Hvitfeldt and Silge (2021) (*https://oreil.ly/k3yCZ*) for more on word embeddings.

Embeddings for a categorical variable can be learned via a TensorFlow neural network with the `step_embed()` function in embed. We can use the outcome alone or optionally the outcome plus a set of additional predictors. Like in feature hashing, the number of new encoding columns to create is a hyperparameter of the feature engineering. We also must make decisions about the neural network structure (the number of hidden units) and how to fit the neural network (how many epochs to train, how much of the data to use for validation in measuring metrics).

Yet one more option available for dealing with a binary outcome is to transform a set of category levels based on their association with the binary outcome. This *weight of evidence* (WoE) transformation (Good 1985) uses the logarithm of the "Bayes factor" (the ratio of the posterior odds to the prior odds) and creates a dictionary mapping each category level to a WoE value. WoE encodings can be determined with the `step_woe()` function in embed.

Chapter Summary

In this chapter, you learned about using preprocessing recipes for encoding categorical predictors. The most straightforward option for transforming a categorical variable to a numeric representation is to create dummy variables from the levels, but this option does not work well when you have a variable with high cardinality (too many levels) or when you may see novel values at prediction time (new levels). One option in such a situation is to create *effect encodings*, a supervised encoding method that uses the outcome. Effect encodings can be learned with or without pooling the categories. Another option uses a *hashing* function to map category levels to a new, smaller set of dummy variables. Feature hashing is fast and has a low-memory footprint. Other options include entity embeddings (learned via a neural network) and weight of evidence transformation.

Most model algorithms require some kind of transformation or encoding of this type for categorical variables. A minority of models, including those based on trees and rules, can handle categorical variables natively and do not require such encodings.

## CHAPTER 18

Explaining Models and Predictions

In Chapter 1, we outlined a taxonomy of models and suggested that models typically are built as one or more of descriptive, inferential, or predictive. We suggested that model performance, as measured by appropriate metrics (like RMSE for regression or area under the ROC curve for classification), can be important for all modeling applications. Similarly, model explanations, answering *why* a model makes the predictions it does, can be important whether the purpose of your model is largely descriptive, to test a hypothesis, or to make a prediction. Answering the question "why?" allows modeling practitioners to understand which features were important in predictions and even how model predictions would change under different values for the features. This chapter covers how to ask a model why it makes the predictions it does.

For some models, like linear regression, it is usually clear how to explain why the model makes its predictions. The structure of a linear model contains coefficients for each predictor that are typically straightforward to interpret. For other models, like random forests that can capture nonlinear behavior by design, it is less transparent how to explain the model's predictions from only the structure of the model itself. Instead, we can apply model explainer algorithms to generate understanding of predictions.

 There are two types of model explanations, *global* and *local*. Global model explanations provide an overall understanding aggregated over a whole set of observations; local model explanations provide information about a prediction for a single observation.

Software for Model Explanations

The tidymodels framework does not itself contain software for model explanations. Instead, models trained and evaluated with tidymodels can be explained with other, supplementary software in R packages such as lime (*https://oreil.ly/bzCAq*), vip (*https://oreil.ly/UpoQf*), and DALEX (*https://oreil.ly/KPZLQ*). We often choose:

- vip functions when we want to use *model-based* methods that take advantage of model structure (and are often faster)

- DALEX functions when we want to use *model-agnostic* methods that can be applied to any model

In Chapters 10 and 11, we trained and compared several models to predict the price of homes in Ames, Iowa, including a linear model with interactions and a random forest model, with results shown in Figure 18-1.

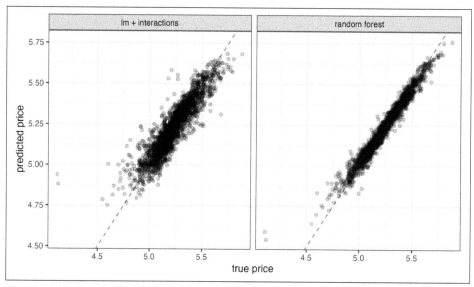

Figure 18-1. Comparing predicted prices for a linear model with interactions and a random forest model.

Let's build model-agnostic explainers for both of these models to find out why they make these predictions. We can use the DALEXtra add-on package for DALEX, which provides support for tidymodels. Biecek and Burzykowski (2021) provide a thorough exploration of how to use DALEX for model explanations; this chapter only summarizes some important approaches, specific to tidymodels. To compute any kind of model explanation, global or local, using DALEX, we first prepare the appropriate data and then create an *explainer* for each model:

```
library(DALEXtra)
vip_features <- c("Neighborhood", "Gr_Liv_Area", "Year_Built",
                  "Bldg_Type", "Latitude", "Longitude")

vip_train <-
  ames_train %>%
  select(all_of(vip_features))

explainer_lm <-
  explain_tidymodels(
    lm_fit,
    data = vip_train,
    y = ames_train$Sale_Price,
    label = "lm + interactions",
    verbose = FALSE
  )

explainer_rf <-
  explain_tidymodels(
    rf_fit,
    data = vip_train,
    y = ames_train$Sale_Price,
    label = "random forest",
    verbose = FALSE
  )
```

 A linear model is typically straightforward to interpret and explain; you may not often find yourself using separate model explanation algorithms for a linear model. However, it can sometimes be difficult to understand or explain the predictions of even a linear model once it has splines and interaction terms!

Dealing with significant feature engineering transformations during model explainability highlights some options we have (or sometimes, ambiguity in such analyses). We can quantify global or local model explanations either in terms of:

- *Original, basic predictors* as they existed without significant feature engineering transformations, or

- *Derived features*, such as those created via dimensionality reduction (Chapter 16) or interactions and spline terms, as in this example

Local Explanations

Local model explanations provide information about a prediction for a single observation. For example, let's consider an older duplex in the North Ames neighborhood (Chapter 4):

```
duplex <- vip_train[120,]
duplex
#> # A tibble: 1 × 6
#>   Neighborhood Gr_Liv_Area Year_Built Bldg_Type Latitude Longitude
```

```
#>    <fct>              <dbl>     <dbl> <fct>     <dbl>   <dbl>
#> 1 North_Ames          1040      1949 Duplex     42.0   -93.6
```

There are multiple possible approaches to understanding why a model predicts a given price for this duplex. One is a break-down explanation, implemented with the DALEX function predict_parts(); it computes how contributions attributed to individual features change the mean model's prediction for a particular observation, like our duplex. For the linear model, the duplex status (Bldg_Type = 3),[1] size, longitude, and age all contribute the most to the price being driven down from the intercept:

```
lm_breakdown <- predict_parts(explainer = explainer_lm, new_observation = duplex)
lm_breakdown
#>                                             contribution
#> lm + interactions: intercept                    5.221
#> lm + interactions: Gr_Liv_Area = 1040          -0.082
#> lm + interactions: Bldg_Type = 3               -0.049
#> lm + interactions: Longitude = -93.608903      -0.043
#> lm + interactions: Year_Built = 1949           -0.039
#> lm + interactions: Latitude = 42.035841        -0.007
#> lm + interactions: Neighborhood = 1             0.001
#> lm + interactions: prediction                   5.002
```

Since this linear model was trained using spline terms for latitude and longitude, the contribution to price for Longitude shown here combines the effects of all of its individual spline terms. The contribution is in terms of the original Longitude feature, not the derived spline features.

The most important features are slightly different for the random forest model, with the size, age, and duplex status being most important:

```
rf_breakdown <- predict_parts(explainer = explainer_rf, new_observation = duplex)
rf_breakdown
#>                                          contribution
#> random forest: intercept                     5.221
#> random forest: Year_Built = 1949            -0.076
#> random forest: Gr_Liv_Area = 1040           -0.075
#> random forest: Bldg_Type = 3                -0.027
#> random forest: Longitude = -93.608903       -0.043
#> random forest: Latitude = 42.035841         -0.028
#> random forest: Neighborhood = 1             -0.003
#> random forest: prediction                    4.969
```

 Model break-down explanations like these depend on the *order* of the features.

---

1 Notice that this package for model explanations focuses on the *level* of categorical predictors in this type of output, like Bldg_Type = 3 for duplex and Neighborhood = 1 for North Ames.

If we choose the `order` for the random forest model explanation to be the same as the default for the linear model (chosen via a heuristic), we can change the relative importance of the features:

```
predict_parts(
  explainer = explainer_rf,
  new_observation = duplex,
  order = lm_breakdown$variable_name
)
#>                                         contribution
#> random forest: intercept                       5.221
#> random forest: Gr_Liv_Area = 1040             -0.075
#> random forest: Bldg_Type = 3                  -0.019
#> random forest: Longitude = -93.608903         -0.023
#> random forest: Year_Built = 1949              -0.104
#> random forest: Latitude = 42.035841           -0.028
#> random forest: Neighborhood = 1               -0.003
#> random forest: prediction                      4.969
```

We can use the fact that these break-down explanations change based on order to compute the most important features over all (or many) possible orderings. This is the idea behind Shapley additive explanations (SHAP) (Lundberg and Lee 2017), where the average contributions of features are computed under different combinations or "coalitions" of feature orderings. Let's compute SHAP attributions for our duplex, using $B = 20$ random orderings:

```
set.seed(1801)
shap_duplex <-
  predict_parts(
    explainer = explainer_rf,
    new_observation = duplex,
    type = "shap",
    B = 20
  )
```

We could use the default plot method from DALEX by calling `plot(shap_duplex)`, or we can access the underlying data and create a custom plot. The box plots in Figure 18-2 display the distribution of contributions across all the orderings we tried, and the bars display the average attribution for each feature:

```
library(forcats)
shap_duplex %>%
  group_by(variable) %>%
  mutate(mean_val = mean(contribution)) %>%
  ungroup() %>%
  mutate(variable = fct_reorder(variable, abs(mean_val))) %>%
  ggplot(aes(contribution, variable, fill = mean_val > 0)) +
  geom_col(data = ~distinct(., variable, mean_val),
           aes(mean_val, variable),
           alpha = 0.5) +
  geom_boxplot(width = 0.5) +
  theme(legend.position = "none") +
  scale_fill_viridis_d() +
  labs(y = NULL)
```

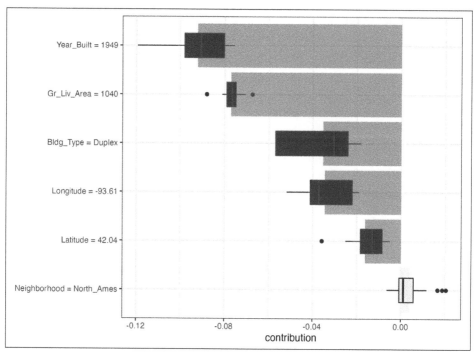

Figure 18-2. Shapley additive explanations from the random forest model for a duplex property.

What about a different observation in our data set? Let's look at a larger, newer one-family home in the Gilbert neighborhood:

```
big_house <- vip_train[1269,]
big_house
#> # A tibble: 1 × 6
#>   Neighborhood Gr_Liv_Area Year_Built Bldg_Type Latitude Longitude
#>   <fct>              <dbl>      <dbl> <fct>        <dbl>     <dbl>
#> 1 Gilbert             2267       2002 OneFam        42.1      -93.6
```

We can compute SHAP average attributions for this house in the same way:

```
set.seed(1802)
shap_house <-
  predict_parts(
    explainer = explainer_rf,
    new_observation = big_house,
    type = "shap",
    B = 20
  )
```

The results are shown in Figure 18-3; unlike the duplex, the size and age of this house contribute to its higher price.

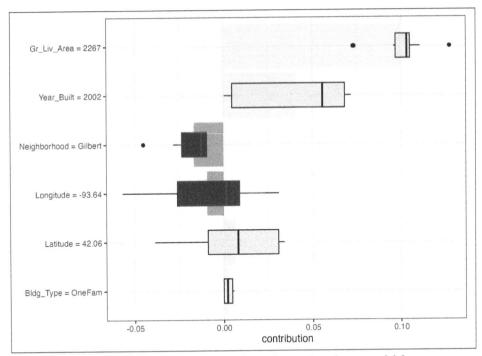

Figure 18-3. Shapley additive explanations from the random forest model for a one-family home in Gilbert.

Global Explanations

Global model explanations, also called global feature importance or variable importance, help us understand which features are most important in driving the predictions of the linear and random forest models overall, aggregated over the whole training set. While the previous section addressed which variables or features are most important in predicting sale price for an individual home, global feature importance addresses the most important variables for a model in aggregate.

 One way to compute variable importance is to *permute* the features (Breiman 2001a). We can permute or shuffle the values of a feature, predict from the model, and then measure how much worse the model fits the data compared to before shuffling.

If shuffling a column causes a large degradation in model performance, it's important; if shuffling a column's values doesn't make much difference to how the model performs, it's not an important variable. This approach can be applied to any kind of model (it is *model agnostic*), and the results are straightforward to understand.

Using DALEX, we compute this kind of variable importance via the `model_parts()` function:

```
set.seed(1803)
vip_lm <- model_parts(explainer_lm, loss_function = loss_root_mean_square)
set.seed(1804)
vip_rf <- model_parts(explainer_rf, loss_function = loss_root_mean_square)
```

Again, we could use the default plot method from DALEX by calling `plot(vip_lm, vip_rf)` but the underlying data is available for exploration, analysis, and plotting. Let's create a function for plotting:

```
ggplot_imp <- function(...) {
  obj <- list(...)
  metric_name <- attr(obj[[1]], "loss_name")
  metric_lab <- paste(metric_name,
                      "after permutations\n(higher indicates more important)")

  full_vip <- bind_rows(obj) %>%
    filter(variable != "_baseline_")

  perm_vals <- full_vip %>%
    filter(variable == "_full_model_") %>%
    group_by(label) %>%
    summarise(dropout_loss = mean(dropout_loss))

  p <- full_vip %>%
    filter(variable != "_full_model_") %>%
    mutate(variable = fct_reorder(variable, dropout_loss)) %>%
    ggplot(aes(dropout_loss, variable))
  if(length(obj) > 1) {
    p <- p +
      facet_wrap(vars(label)) +
      geom_vline(data = perm_vals, aes(xintercept = dropout_loss, color = label),
                 size = 1.4, lty = 2, alpha = 0.7) +
      geom_boxplot(aes(color = label, fill = label), alpha = 0.2)
  } else {
    p <- p +
      geom_vline(data = perm_vals, aes(xintercept = dropout_loss),
                 size = 1.4, lty = 2, alpha = 0.7) +
      geom_boxplot(fill = "#91CBD765", alpha = 0.4)

  }
  p +
    theme(legend.position = "none") +
    labs(x = metric_lab,
         y = NULL,  fill = NULL,  color = NULL)
}
```

Using `ggplot_imp(vip_lm, vip_rf)` produces Figure 18-4.

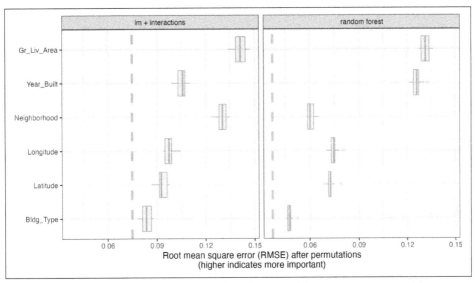

Figure 18-4. Global explainer for the random forest and linear regression models.

The dashed line in each panel of Figure 18-4 shows the RMSE for the full model, either the linear model or the random forest model. Features farther to the right are more important because permuting them results in higher RMSE. There is quite a lot of interesting information to learn from this plot; for example, neighborhood is quite important in the linear model with interactions/splines but the second least important feature for the random forest model.

Building Global Explanations from Local Explanations

So far in this chapter, we have focused on local model explanations for a single observation (via Shapley additive explanations) and global model explanations for a data set as a whole (via permuting features). It is also possible to build global model explanations by aggregating local model explanations, as with *partial dependence profiles*.

Partial dependence profiles show how the expected value of a model prediction, like the predicted price of a home in Ames, changes as a function of a feature, like the age or gross living area.

One way to build such a profile is by aggregating or averaging profiles for individual observations. A profile showing how an individual observation's prediction changes as a function of a given feature is called an ICE (individual conditional expectation) profile or a CP (ceteris paribus) profile. We can compute such individual profiles (for 500 of the observations in our training set) and then aggregate them using the DALEX function model_profile():

```
set.seed(1805)
pdp_age <- model_profile(explainer_rf, N = 500, variables = "Year_Built")
```

Let's create another function for plotting the underlying data in this object:

```
ggplot_pdp <- function(obj, x) {

  p <-
    as_tibble(obj$agr_profiles) %>%
    mutate(`_label_` = stringr::str_remove(`_label_`, "^[^_]*_")) %>%
    ggplot(aes(`_x_`, `_yhat_`)) +
    geom_line(data = as_tibble(obj$cp_profiles),
              aes(x = {{ x }}, group = `_ids_`),
              size = 0.5, alpha = 0.05, color = "gray50")

  num_colors <- n_distinct(obj$agr_profiles$`_label_`)

  if (num_colors > 1) {
    p <- p + geom_line(aes(color = `_label_`, lty = `_label_`), size = 1.2)
  } else {
    p <- p + geom_line(color = "midnightblue", size = 1.2, alpha = 0.8)
  }

  p
}
```

Using this function generates Figure 18-5, where we can see the nonlinear behavior of the random forest model:

```
ggplot_pdp(pdp_age, Year_Built)  +
  labs(x = "Year built",
       y = "Sale Price (log)",
       color = NULL)
```

Sale price for houses built in different years is mostly flat, with a modest rise after about 1960. Partial dependence profiles can be computed for any other feature in the model, and also for groups in the data, such as Bldg_Type. Let's use 1,000 observations for these profiles:

```
set.seed(1806)
pdp_liv <- model_profile(explainer_rf, N = 1000,
                         variables = "Gr_Liv_Area",
                         groups = "Bldg_Type")

ggplot_pdp(pdp_liv, Gr_Liv_Area) +
  scale_x_log10() +
  scale_color_brewer(palette = "Dark2") +
  labs(x = "Gross living area",
       y = "Sale Price (log)",
       color = NULL, lty = NULL)
```

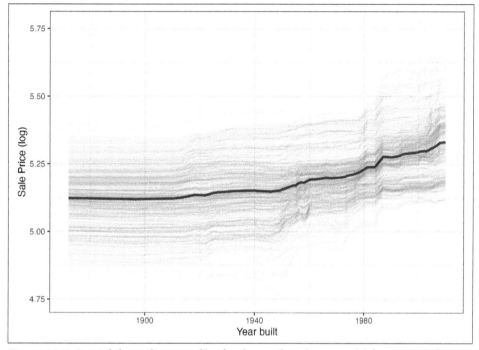

Figure 18-5. Partial dependence profiles for the random forest model focusing on the year built predictor.

This code produces Figure 18-6, where we see that sale price increases the most between about 1,000 and 3,000 square feet of living area, and that different home types (like single family homes or different types of townhouses) mostly exhibit similar increasing trends in price with more living space.

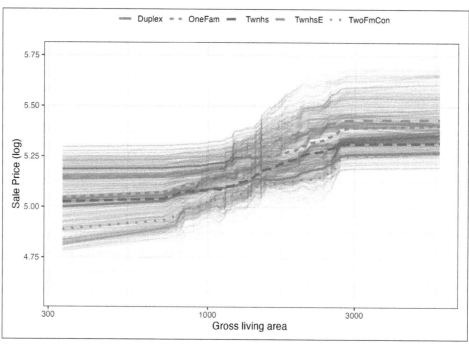

Figure 18-6. Partial dependence profiles for the random forest model focusing on building types and gross living area.

We have the option of using `plot(pdp_liv)` for default DALEX plots, but since we are making plots with the underlying data here, we can even facet by one of the features to visualize if the predictions change differently and highlighting the imbalance in these subgroups (as shown in Figure 18-7):

```
as_tibble(pdp_liv$agr_profiles) %>%
  mutate(Bldg_Type = stringr::str_remove(`_label_`, "random forest_")) %>%
  ggplot(aes(`_x_`, `_yhat_`, color = Bldg_Type)) +
  geom_line(data = as_tibble(pdp_liv$cp_profiles),
            aes(x = Gr_Liv_Area, group = `_ids_`),
            size = 0.5, alpha = 0.1, color = "gray50") +
  geom_line(size = 1.2, alpha = 0.8, show.legend = FALSE) +
  scale_x_log10() +
  facet_wrap(~Bldg_Type) +
  scale_color_brewer(palette = "Dark2") +
  labs(x = "Gross living area",
       y = "Sale Price (log)",
       color = NULL)
```

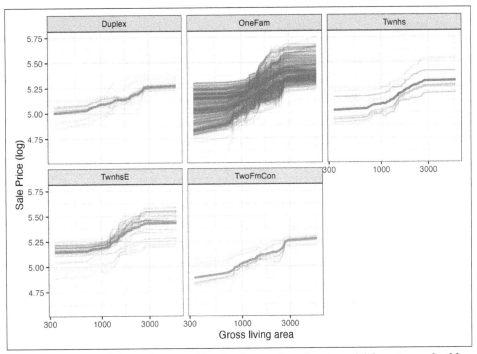

Figure 18-7. Partial dependence profiles for the random forest model focusing on building types and gross living area using facets.

There is no one correct approach for building model explanations, and the options outlined in this chapter are not exhaustive. We have highlighted good options for explanations at both the individual and global level, as well as how to bridge from one to the other, and we point you to Biecek and Burzykowski (2021) and Molnar (2020) (*https://oreil.ly/P5Vlg*) for further reading.

Back to Beans!

In Chapter 16, we discussed how to use dimensionality reduction as a feature engineering or preprocessing step when modeling high-dimensional data. For our example data set of dry bean morphology measures predicting bean type, we saw great results from partial least squares (PLS) dimensionality reduction combined with a regularized discriminant analysis model. Which of those morphological characteristics were *most* important in the bean type predictions? We can use the same approach outlined throughout this chapter to create a model-agnostic explainer and compute, say, global model explanations via model_parts():

```
set.seed(1807)
vip_beans <-
  explain_tidymodels(
    rda_wflow_fit,
    data = bean_train %>% select(-class),
    y = bean_train$class,
    label = "RDA",
    verbose = FALSE
  ) %>%
  model_parts()
```

Using our previously defined importance plotting function, `ggplot_imp(vip_beans)` produces Figure 18-8.

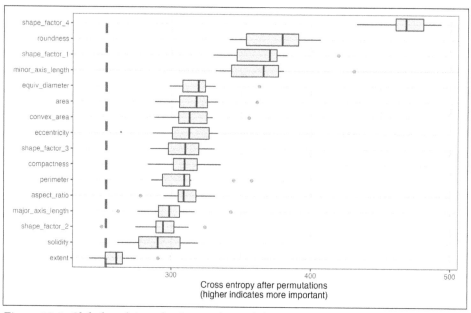

Figure 18-8. Global explainer for the regularized discriminant analysis model on the beans data.

 The measures of global feature importance that we see in Figure 18-8 incorporate the effects of all of the PLS components but in terms of the original variables.

Figure 18-8 shows us that shape factors are among the most important features for predicting bean type, especially shape factor 4, a measure of solidity that takes into account the area A, major axis L, and minor axis l:

$$SF4 = \frac{A}{\pi(L/2)(l/2)}$$

We can see from Figure 18-8 that shape factor 1 (the ratio of the major axis to the area), the minor axis length, and roundness are the next most important bean characteristics for predicting bean variety.

Chapter Summary

For some types of models, the answer to why a model made a certain prediction is straightforward, but for other types of models, we must use separate explainer algorithms to understand what features are relatively most important for predictions. You can generate two main kinds of model explanations from a trained model. Global explanations provide information aggregated over an entire data set, while local explanations provide understanding about a model's predictions for a single observation.

Packages such as DALEX and its supporting package DALEXtra, vip, and lime can be integrated into a tidymodels analysis to provide these model explainers. Model explanations are just one piece of understanding whether your model is appropriate and effective, along with estimates of model performance; Chapter 19 further explores the quality and trustworthiness of predictions.

When Should You Trust Your Predictions?

A predictive model can almost always produce a prediction, given input data. However, in plenty of situations it is inappropriate to produce such a prediction. When a new data point is well outside of the range of data used to create the model, making a prediction may be an inappropriate *extrapolation*. A more qualitative example of an inappropriate prediction would be when the model is used in a completely different context. The cell segmentation data used in Chapter 14 flags when human breast cancer cells can or cannot be accurately isolated inside an image. A model built from these data could be inappropriately applied to stomach cells for the same purpose. We can produce a prediction, but it is unlikely to be applicable to the different cell type.

This chapter discusses two methods for quantifying the potential prediction quality:

Equivocal zones
> This method uses the predicted values to alert the user that results may be suspect.

Applicability
> This method uses the predictors to measure the amount of extrapolation (if any) for new samples.

Equivocal Results

If a model result indicated that you had a 51% chance of having contracted COVID-19, it would be natural to view the diagnosis with some skepticism. In fact, regulatory bodies often require many medical diagnostics to have an *equivocal zone*. This zone is a range of results in which the prediction should not be reported to patients, for example, some range of COVID-19 test results that are too uncertain to be reported to a patient. See Danowski et al. (1970) and Kerleguer et al. (2003)

for examples. The same notion can be applied to models created outside of medical diagnostics.

 In some cases, the amount of uncertainty associated with a prediction is too high to be trusted.

Let's use a function that can simulate classification data with two classes and two predictors (x and y). The true model is a logistic regression model with the equation:

$$\text{logit}(p) = -1 - 2x - \frac{x^2}{5} + 2y^2$$

The two predictors follow a bivariate normal distribution with a correlation of 0.70. We'll create a training set of 200 samples and a test set of 50:

```
library(tidymodels)
tidymodels_prefer()

simulate_two_classes <-
  function (n, error = 0.1, eqn = quote(-1 - 2 * x - 0.2 * x^2 + 2 * y^2))  {
    # Slightly correlated predictors
    sigma <- matrix(c(1, 0.7, 0.7, 1), nrow = 2, ncol = 2)
    dat <- MASS::mvrnorm(n = n, mu = c(0, 0), Sigma = sigma)
    colnames(dat) <- c("x", "y")
    cls <- paste0("class_", 1:2)
    dat <-
      as_tibble(dat) %>%
      mutate(
        linear_pred = !!eqn,
        # Add some misclassification noise
        linear_pred = linear_pred + rnorm(n, sd = error),
        prob = binomial()$linkinv(linear_pred),
        class = ifelse(prob > runif(n), cls[1], cls[2]),
        class = factor(class, levels = cls)
      )
    dplyr::select(dat, x, y, class)
  }

set.seed(1901)
training_set <- simulate_two_classes(200)
testing_set  <- simulate_two_classes(50)
```

We estimate a logistic regression model using Bayesian methods (using the default Gaussian prior distributions for the parameters):

```
two_class_mod <-
  logistic_reg() %>%
  set_engine("stan", seed = 1902) %>%
  fit(class ~ . + I(x^2)+ I(y^2), data = training_set)
print(two_class_mod, digits = 3)
#> parsnip model object
#>
#> stan_glm
#>  family:       binomial [logit]
#>  formula:      class ~ . + I(x^2) + I(y^2)
#>  observations: 200
#>  predictors:   5
#> ------
#>             Median MAD_SD
#> (Intercept)  1.092  0.287
#> x            2.290  0.423
#> y            0.314  0.354
#> I(x^2)       0.077  0.307
#> I(y^2)      -2.465  0.424
#>
#> ------
#> * For help interpreting the printed output see ?print.stanreg
#> * For info on the priors used see ?prior_summary.stanreg
```

The fitted class boundary is overlaid onto the test set in Figure 19-1. The data points closest to the class boundary are the most uncertain. If their values changed slightly, their predicted class might change. One simple method for disqualifying some results is to call them "equivocal" if the values are within some range around 50% (or the appropriate probability cutoff for a certain situation). Depending on the problem the model is being applied to, this might indicate we should collect another measurement or require more information before a trustworthy prediction is possible.

We could base the width of the band around the cutoff on how performance improves when the uncertain results are removed. However, we should also estimate the reportable rate (the expected proportion of usable results). For example, it would not be useful in real-world situations to have perfect performance but release predictions on only 2% of the samples passed to the model.

Let's use the test set to determine the balance between improving performance and having enough reportable results. The predictions are created using:

```
test_pred <- augment(two_class_mod, testing_set)
test_pred %>% head()
#> # A tibble: 6 × 6
#>        x      y class   .pred_class .pred_class_1 .pred_class_2
#>    <dbl>  <dbl> <fct>   <fct>               <dbl>         <dbl>
#> 1  1.12  -0.176 class_2 class_2            0.0256         0.974
#> 2 -0.126 -0.582 class_2 class_1            0.555          0.445
#> 3  1.92   0.615 class_2 class_2            0.00620        0.994
#> 4 -0.400  0.252 class_2 class_2            0.472          0.528
#> 5  1.30   1.09  class_1 class_2            0.163          0.837
#> 6  2.59   1.36  class_2 class_2            0.0317         0.968
```

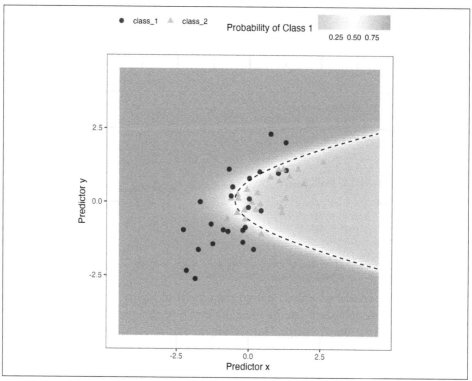

Figure 19-1. Simulated two-class data set with a logistic regression fit and decision boundary.

With tidymodels, the probably package contains functions for equivocal zones. For cases with two classes, the `make_two_class_pred()` function creates a factor-like column that has the predicted classes with an equivocal zone:

```
library(probably)

lvls <- levels(training_set$class)

test_pred <-
  test_pred %>%
  mutate(.pred_with_eqz = make_two_class_pred(.pred_class_1, lvls, buffer = 0.15))

test_pred %>% count(.pred_with_eqz)
#> # A tibble: 3 × 2
#>   .pred_with_eqz     n
#>       <clss_prd> <int>
#> 1           [EQ]     9
#> 2        class_1    20
#> 3        class_2    21
```

Rows that are within 0.50 ± 0.15 are given a value of [EQ].

Important: [EQ] in this example is not a factor level but an attribute of that column.

Since the factor levels are the same as the original data, confusion matrices and other statistics can be computed without error. When using standard functions from the yardstick package, the equivocal results are converted to NA and are not used in the calculations that use the hard class predictions. Notice the differences in these confusion matrices:

```
# All data
test_pred %>% conf_mat(class, .pred_class)
#>           Truth
#> Prediction class_1 class_2
#>    class_1    20       6
#>    class_2     5      19

# Reportable results only:
test_pred %>% conf_mat(class, .pred_with_eqz)
#>           Truth
#> Prediction class_1 class_2
#>    class_1    17       3
#>    class_2     5      16
```

An is_equivocal() function is also available for filtering these rows from the data.

Does the equivocal zone help improve accuracy? Let's look at different buffer sizes, as shown in Figure 19-2:

```
# A function to change the buffer then compute performance.
eq_zone_results <- function(buffer) {
  test_pred <-
    test_pred %>%
    mutate(.pred_with_eqz = make_two_class_pred(.pred_class_1, lvls, buffer = buffer))
  acc <- test_pred %>% accuracy(class, .pred_with_eqz)
  rep_rate <- reportable_rate(test_pred$.pred_with_eqz)
  tibble(accuracy = acc$.estimate, reportable = rep_rate, buffer = buffer)
}

# Evaluate a sequence of buffers and plot the results.
map_dfr(seq(0, .1, length.out = 40), eq_zone_results) %>%
  pivot_longer(c(-buffer), names_to = "statistic", values_to = "value") %>%
  ggplot(aes(x = buffer, y = value, lty = statistic)) +
  geom_step(size = 1.2, alpha = 0.8) +
  labs(y = NULL, lty = NULL)
```

Figure 19-2 shows us that accuracy improves by a few percentage points but at the cost of nearly 10% of predictions being unusable! The value of such a compromise depends on how the model predictions will be used.

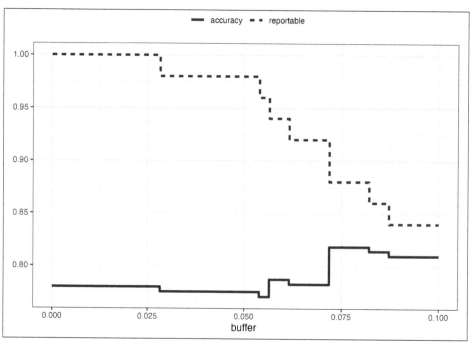

Figure 19-2. The effect of equivocal zones on model performance.

This analysis focused on using the predicted class probability to disqualify points, since this is a fundamental measure of uncertainty in classification models. A slightly better approach would be to use the standard error of the class probability. Since we used a Bayesian model, the probability estimates we found are actually the mean of the posterior predictive distribution. In other words, the Bayesian model gives us a distribution for the class probability. Measuring the standard deviation of this distribution gives us a *standard error of prediction* of the probability. In most cases, this value is directly related to the mean class probability. You might recall that, for a Bernoulli random variable with probability p, the variance is $p(1 - p)$. Because of this relationship, the standard error is largest when the probability is 50%. Instead of assigning an equivocal result using the class probability, we could instead use a cutoff on the standard error of prediction.

One important aspect of the standard error of prediction is that it takes into account more than just the class probability. In cases where there is significant extrapolation or aberrant predictor values, the standard error might increase. The benefit of using the standard error of prediction is that it might also flag predictions that are problematic (as opposed to simply uncertain). One reason we used the Bayesian model is that it naturally estimates the standard error of prediction; not many models can calculate this. For our test set, using type = "pred_int" will produce upper and lower limits and the std_error adds a column for that quantity. For 80% intervals:

```
test_pred <-
  test_pred %>%
  bind_cols(
    predict(two_class_mod, testing_set, type = "pred_int", std_error = TRUE)
  )
```

For our example where the model and data are well behaved, Figure 19-3 shows the standard error of prediction across the space:

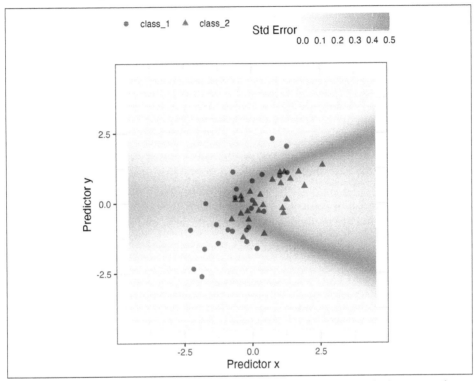

Figure 19-3. The effect of the standard error of prediction overlaid with the test set data.

Using the standard error as a measure to preclude samples from being predicted can also be applied to models with numeric outcomes. However, as shown in the next section, this may not always work.

Determining Model Applicability

Equivocal zones try to measure the reliability of a prediction based on the model outputs. It may be that model statistics, such as the standard error of prediction, cannot measure the impact of extrapolation, and so we need another way to assess whether to trust a prediction and answer, "Is our model applicable for predicting a specific data point?" Let's take the Chicago train data used extensively in Kuhn and

Johnson (2020) (*https://oreil.ly/rsZqK*) and first shown in Chapter 2. The goal is to predict the number of customers entering the Clark and Lake train station each day.

The data set in the modeldata package (a tidymodels package with example data sets) has daily values between January 22, 2001, and August 28, 2016. Let's create a small test set using the last two weeks of the data:

```
## loads both `Chicago` dataset as well as `stations`
data(Chicago)

Chicago <- Chicago %>% select(ridership, date, one_of(stations))

n <- nrow(Chicago)

Chicago_train <- Chicago %>% slice(1:(n - 14))
Chicago_test  <- Chicago %>% slice((n - 13):n)
```

The main predictors are lagged ridership data at different train stations, including Clark and Lake, as well as the date. The ridership predictors are highly correlated with one another. In the following recipe, the date column is expanded into several new features, and the ridership predictors are represented using partial least squares (PLS) components. PLS (Geladi and Kowalski 1986), as we discussed in Chapter 16, is a supervised version of principal component analysis where the new features have been decorrelated but are predictive of the outcome data.

Using the preprocessed data, we fit a standard linear model:

```
base_recipe <-
  recipe(ridership ~ ., data = Chicago_train) %>%
  # Create date features
  step_date(date) %>%
  step_holiday(date, keep_original_cols = FALSE) %>%
  # Create dummy variables from factor columns
  step_dummy(all_nominal()) %>%
  # Remove any columns with a single unique value
  step_zv(all_predictors()) %>%
  step_normalize(!!!stations)%>%
  step_pls(!!!stations, num_comp = 10, outcome = vars(ridership))

lm_spec <-
  linear_reg() %>%
  set_engine("lm")

lm_wflow <-
  workflow() %>%
  add_recipe(base_recipe) %>%
  add_model(lm_spec)

set.seed(1902)
lm_fit <- fit(lm_wflow, data = Chicago_train)
```

How well do the data fit on the test set? We can `predict()` for the test set to find both predictions and prediction intervals:

```
res_test <-
  predict(lm_fit, Chicago_test) %>%
```

```
  bind_cols(
    predict(lm_fit, Chicago_test, type = "pred_int"),
    Chicago_test
  )

res_test %>% select(date, ridership, starts_with(".pred"))
#> # A tibble: 14 × 5
#>   date         ridership .pred .pred_lower .pred_upper
#>   <date>           <dbl> <dbl>       <dbl>       <dbl>
#> 1 2016-08-15        20.6  20.3        16.2        24.5
#> 2 2016-08-16        21.0  21.3        17.1        25.4
#> 3 2016-08-17        21.0  21.4        17.3        25.6
#> 4 2016-08-18        21.3  21.4        17.3        25.5
#> 5 2016-08-19        20.4  20.9        16.7        25.0
#> 6 2016-08-20        6.22  7.52        3.34        11.7
#> # … with 8 more rows
res_test %>% rmse(ridership, .pred)
#> # A tibble: 1 × 3
#>   .metric .estimator .estimate
#>   <chr>   <chr>          <dbl>
#> 1 rmse    standard       0.865
```

These are fairly good results. Figure 19-4 visualizes the predictions along with 95% prediction intervals.

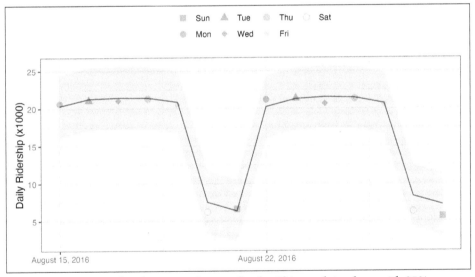

Figure 19-4. Two weeks of 2016 predictions for the Chicago data along with 95% prediction intervals.

Given the scale of the ridership numbers, these results look particularly good for such a simple model. If this model were deployed, how well would it have done a few years later in June 2020? The model successfully makes a prediction, as a predictive model almost always will when given input data:

```
res_2020 <-
  predict(lm_fit, Chicago_2020) %>%
  bind_cols(
    predict(lm_fit, Chicago_2020, type = "pred_int"),
    Chicago_2020
  )

res_2020 %>% select(date, contains(".pred"))
#> # A tibble: 14 × 4
#>    date        .pred .pred_lower .pred_upper
#>    <date>      <dbl>       <dbl>       <dbl>
#> 1 2020-06-01   20.1        15.9        24.3
#> 2 2020-06-02   21.4        17.2        25.6
#> 3 2020-06-03   21.5        17.3        25.6
#> 4 2020-06-04   21.3        17.1        25.4
#> 5 2020-06-05   20.7        16.6        24.9
#> 6 2020-06-06   9.04         4.88       13.2
#> # … with 8 more rows
```

The prediction intervals are about the same width, even though these data are well beyond the time period of the original training set. However, given the global pandemic in 2020, the performance on these data is abysmal:

```
res_2020 %>% select(date, ridership, starts_with(".pred"))
#> # A tibble: 14 × 5
#>    date       ridership .pred .pred_lower .pred_upper
#>    <date>         <dbl> <dbl>       <dbl>       <dbl>
#> 1 2020-06-01     0.002  20.1        15.9        24.3
#> 2 2020-06-02     0.005  21.4        17.2        25.6
#> 3 2020-06-03     0.566  21.5        17.3        25.6
#> 4 2020-06-04     1.66   21.3        17.1        25.4
#> 5 2020-06-05     1.95   20.7        16.6        24.9
#> 6 2020-06-06     1.08    9.04        4.88       13.2
#> # … with 8 more rows
res_2020 %>% rmse(ridership, .pred)
#> # A tibble: 1 × 3
#>    .metric .estimator .estimate
#>    <chr>   <chr>          <dbl>
#> 1 rmse     standard        17.2
```

You can see this terrible model performance visually in Figure 19-5.

Confidence and prediction intervals for linear regression expand as the data become more and more removed from the center of the training set. However, that effect is not dramatic enough to flag these predictions as being poor.

 Sometimes the statistics produced by models don't measure the quality of predictions very well.

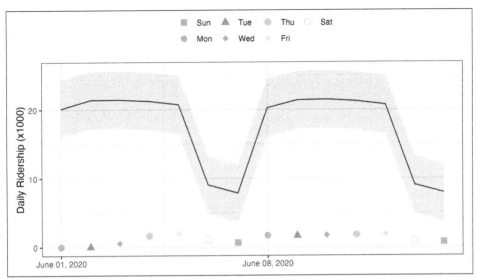

Figure 19-5. Two weeks of 2020 predictions for the Chicago data along with 95% prediction intervals.

This situation can be avoided by having a secondary methodology that can quantify how applicable the model is for any new prediction (i.e., the model's *applicability domain*). There are a variety of methods to compute an applicability domain model, such as Jaworska, Nikolova-Jeliazkova, and Aldenberg (2005) or Netzeva et al. (2005). The approach used in this chapter is a fairly simple unsupervised method that attempts to measure how much (if any) a new data point is beyond the training data.[1]

The idea is to accompany a prediction with a score that measures how similar the new point is to the training set.

One method that works well uses principal component analysis (PCA) on the numeric predictor values. We'll illustrate the process by using only two of the predictors that correspond to ridership at different stations (California and Austin stations). The training set are shown in panel (a) in Figure 19-6. The ridership data for these stations are highly correlated, and the two distributions shown in the scatter plot correspond to ridership on the weekends and weekdays.

---

1 Bartley et al. (2019) show yet another method and apply it to ecological studies.

The first step is to conduct PCA on the training data. The PCA scores for the training set are shown in panel (b) in Figure 19-6. Next, using these results, we measure the distance of each training set point to the center of the PCA data (panel (c) of Figure 19-6). We can then use this *reference distribution* (panel (d) of Figure 19-6) to estimate how far a data point is from the mainstream of the training data.

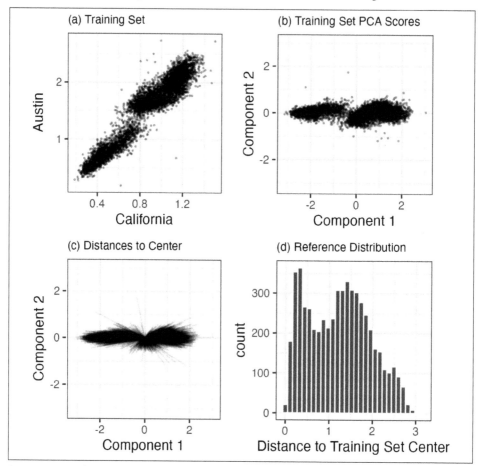

Figure 19-6. The PCA reference distribution based on the training set.

For a new sample, the PCA scores are computed along with the distance to the center of the training set.

However, what does it mean when a new sample has a distance of X? Since the PCA components can have different ranges from data set to data set, there is no obvious limit to say that a distance is too large.

One approach is to treat the distances from the training set data as "normal." For new samples, we can determine how the new distance compares to the range in the reference distribution (from the training set). A percentile can be computed for new samples that reflect how much of the training set is less extreme than the new samples.

 A percentile of 90% means that most of the training set data are closer to the data center than the new sample.

The plot in Figure 19-7 overlays a testing set sample (triangle and dashed line) and a 2020 sample (circle and solid line) with the PCA distances from the training set.

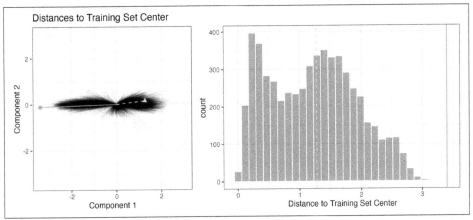

Figure 19-7. The reference distribution with two new points: one using the test set and one from the 2020 data.

The test set point has a distance of 1.28. It is in the 51.8% percentile of the training set distribution, indicating that it is snugly within the mainstream of the training set.

The 2020 sample is farther from the center than any of the training set samples (with a percentile of 100%). This indicates the sample is very extreme and that its corresponding prediction would be a severe extrapolation (and probably should not be reported).

The applicable package can develop an applicability domain model using PCA. We'll use the 20 lagged station ridership predictors as inputs into the PCA analysis. There is an additional argument called threshold that determines how many components are used in the distance calculation. For our example, we'll use a large value that indicates we should use enough components to account for 99% of the variation in the ridership predictors:

```
library(applicable)
pca_stat <- apd_pca(~ ., data = Chicago_train %>% select(one_of(stations)),
                    threshold = 0.99)
pca_stat
#> # Predictors:
#>    20
#> # Principal Components:
#>    9 components were needed
#>    to capture at least 99% of the
#>    total variation in the predictors.
```

The `autoplot()` method plots the reference distribution. It has an optional argument for which data to plot. We'll add a value of `distance` to plot only the training set distance distribution. This code generates the plot in Figure 19-8:

```
autoplot(pca_stat, distance) + labs(x = "distance")
```

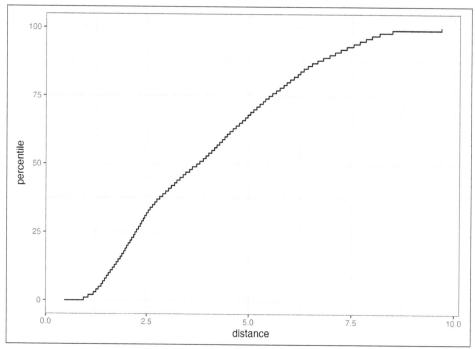

Figure 19-8. The results of using the `autoplot()` method on an applicable object.

The x-axis shows the values of the distance and the y-axis displays the distribution's percentiles. For example, half of the training set samples had distances less than 3.7.

To compute the percentiles for new data, the `score()` function works in the same way as `predict()`:

```
score(pca_stat, Chicago_test) %>% select(starts_with("distance"))
#> # A tibble: 14 × 2
#>    distance distance_pctl
#>       <dbl>         <dbl>
```

```
#> 1    4.88      66.7
#> 2    5.21      71.4
#> 3    5.19      71.1
#> 4    5.00      68.5
#> 5    4.36      59.3
#> 6    4.10      55.2
#> # … with 8 more rows
```

These seem fairly reasonable. For the 2020 data:

```
score(pca_stat, Chicago_2020) %>% select(starts_with("distance"))
#> # A tibble: 14 × 2
#>    distance distance_pctl
#>       <dbl>         <dbl>
#> 1     9.39          99.8
#> 2     9.40          99.8
#> 3     9.30          99.7
#> 4     9.30          99.7
#> 5     9.29          99.7
#> 6    10.1              1
#> # … with 8 more rows
```

The 2020 distance values indicate that these predictor values are outside of the vast majority of data seen by the model at training time. These should be flagged so that the predictions are either not reported at all or viewed with skepticism.

 One important aspect of this analysis concerns which predictors are used to develop the applicability domain model. In our analysis, we used the raw predictor columns. However, in building the model, PLS score features were used in their place. Which of these should apd_pca() use? The apd_pca() function can also take a recipe as the input (instead of a formula) so that the distances reflect the PLS scores instead of the individual predictor columns. You can evaluate both methods to understand which one gives more relevant results.

Chapter Summary

This chapter showed two methods for evaluating whether predictions should be reported to the consumers of models. Equivocal zones deal with outcomes/predictions and can be helpful when the amount of uncertainty in a prediction is too large.

Applicability domain models deal with features/predictors and quantify the amount of extrapolation (if any) that occurs when making a prediction. This chapter showed a basic method using principal component analysis, although there are many other ways to measure applicability. The applicable package also contains specialized methods for data sets where all of the predictors are binary. This method computes similarity scores between training set data points to define the reference distribution.

## CHAPTER 20

Ensembles of Models

A model ensemble, where the predictions of multiple single learners are aggregated to make one prediction, can produce a high-performance final model. The most popular methods for creating ensemble models are bagging (Breiman 1996a), random forest (Ho 1995; Breiman 2001a), and boosting (Freund and Schapire 1997). Each of these methods combines the predictions from multiple versions of the same type of model (e.g., classification trees). However, one of the earliest methods for creating ensembles is *model stacking* (Wolpert 1992; Breiman 1996b).

Model stacking combines the predictions for multiple models of any type. For example, a logistic regression, a classification tree, and a support vector machine can be included in a stacking ensemble.

This chapter shows how to stack predictive models using the stacks package. We'll reuse the results from Chapter 15 where multiple models were evaluated to predict the compressive strength of concrete mixtures.

The process of building a stacked ensemble is:

1. Assemble the training set of holdout predictions (produced via resampling).

2. Create a model to blend these predictions.

3. For each member of the ensemble, fit the model on the original training set.

In subsequent sections, we'll describe this process. However, before proceeding, we'll clarify some nomenclature for the variations of what "the model" can mean. This can quickly become an overloaded term when we are working on a complex modeling analysis! Let's consider the multilayer perceptron (MLP) model (a.k.a. neural network) created in Chapter 15.

In general, we'll talk about an MLP model as the *type* of model. Linear regression and support vector machines are other model types.

Tuning parameters are an important aspect of a model. Back in Chapter 15, the MLP model was tuned over 25 tuning parameter values. In the previous chapters, we've called these *candidate tuning parameter* values or *model configurations*. In literature on ensembling these have also been called the base models.

We'll use the term *candidate members* to describe the possible model configurations (of all model types) that might be included in the stacking ensemble.

This means that a stacking model can include different types of models (e.g., trees and neural networks) and different configurations of the same model (e.g., trees with different depths).

Creating the Training Set for Stacking

The first step for building a stacked ensemble relies on the assessment set predictions from a resampling scheme with multiple splits. For each data point in the training set, stacking requires an out-of-sample prediction of some sort. For regression models, this is the predicted outcome. For classification models, the predicted classes or probabilities are available for use, although the latter contains more information than the hard class predictions. For a set of models, a data set is assembled where rows are the training set samples and columns are the out-of-sample predictions from the set of multiple models.

Back in Chapter 15, we used five repeats of 10-fold cross-validation to resample the data. This resampling scheme generates five assessment set predictions for each training set sample. Multiple out-of-sample predictions can occur in several other resampling techniques (e.g., bootstrapping). For the purpose of stacking, any replicate predictions for a data point in the training set are averaged so that there is a single prediction per training set sample per candidate member.

Simple validation sets can also be used with stacking since tidymodels considers this to be a single resample.

For the concrete example, the training set used for model stacking has columns for all of the candidate tuning parameter results. Table 20-1 presents the first six rows and selected columns.

Table 20-1. Predictions from candidate tuning parameter configurations

Sample #	Bagged tree	MARS 1	MARS 2	Cubist 1	...	Cubist 25	...
1	25.18	17.92	17.21	17.79		17.82	
2	5.18	-1.77	-0.74	2.83		3.87	
3	9.71	7.26	5.91	6.31		8.60	
4	25.21	20.93	21.52	23.72		21.61	
5	6.33	1.53	0.14	3.60		4.57	
6	7.88	4.88	1.74	7.69		7.55	

There is a single column for the bagged tree model since it has no tuning parameters. Also, recall that MARS was tuned over a single parameter (the product degree) with two possible configurations, so this model is represented by two columns. Most of the other models have 25 corresponding columns, as shown for Cubist in this example.

For classification models, the candidate prediction columns would be predicted class probabilities. Since these columns add to one for each model, the probabilities for one of the classes can be left out.

To summarize where we are so far, the first step to stacking is to assemble the assessment set predictions for the training set from each candidate model. We can use these assessment set predictions to move forward and build a stacked ensemble.

To start ensembling with the stacks package, create an empty data stack using the `stacks()` function and then add candidate models. Recall that we used workflow sets to fit a wide variety of models to these data. We'll use the racing results:

```
race_results
#> # A workflow set/tibble: 12 × 4
#>   wflow_id     info             option      result
#>   <chr>        <list>           <list>      <list>
#> 1 MARS         <tibble [1 × 4]> <opts[3]>   <race[+]>
#> 2 CART         <tibble [1 × 4]> <opts[3]>   <race[+]>
#> 3 CART_bagged  <tibble [1 × 4]> <opts[3]>   <rsmp[+]>
#> 4 RF           <tibble [1 × 4]> <opts[3]>   <race[+]>
#> 5 boosting     <tibble [1 × 4]> <opts[3]>   <race[+]>
#> 6 Cubist       <tibble [1 × 4]> <opts[3]>   <race[+]>
#> # … with 6 more rows
```

In this case, our syntax is:

```
library(tidymodels)
library(stacks)
tidymodels_prefer()

concrete_stack <-
  stacks() %>%
  add_candidates(race_results)

concrete_stack
#> # A data stack with 12 model definitions and 21 candidate members:
#> #   MARS: 1 model configuration
#> #   CART: 1 model configuration
#> #   CART_bagged: 1 model configuration
#> #   RF: 1 model configuration
#> #   boosting: 1 model configuration
#> #   Cubist: 1 model configuration
#> #   SVM_radial: 1 model configuration
#> #   SVM_poly: 1 model configuration
#> #   KNN: 3 model configurations
#> #   neural_network: 4 model configurations
#> #   full_quad_linear_reg: 5 model configurations
#> #   full_quad_KNN: 1 model configuration
#> # Outcome: compressive_strength (numeric)
```

Recall that racing methods (introduced in Chapter 13) are more efficient since they might not evaluate all configurations on all resamples. Stacking requires that all candidate members have the complete set of resamples. add_candidates() includes only the model configurations that have complete results.

Why use the racing results instead of the full set of candidate models contained in grid_results? Either can be used. We found better performance for these data using the racing results. This might be due to the racing method preselecting the best model(s) from the larger grid.

If we had not used the workflowsets package, objects from the tune and finetune could also be passed to add_candidates(). This can include both grid and iterative search objects.

Blend the Predictions

The training set predictions and the corresponding observed outcome data are used to create a *meta-learning model* where the assessment set predictions are the predictors of the observed outcome data. Meta-learning can be accomplished using any model. The most commonly used model is a regularized generalized linear model, which encompasses linear, logistic, and multinomial models. Specifically, regularization via the lasso penalty (Tibshirani 1996), which uses shrinkage to pull points toward a central value, has several advantages:

- Using the lasso penalty can remove candidates (and sometimes whole model types) from the ensemble.
- The correlation between ensemble candidates tends to be very high, and regularization helps alleviate this issue.

Breiman (1996b) also suggested that, when a linear model is used to blend the predictions, it might be helpful to constrain the blending coefficients to be nonnegative. We have generally found this to be good advice, and it is the default for the stacks package (but it can be changed via an optional argument).

Since our outcome is numeric, linear regression is used for the metamodel. Fitting the metamodel is as straightforward as using:

```
set.seed(2001)
ens <- blend_predictions(concrete_stack)
```

This evaluates the meta-learning model over a predefined grid of lasso penalty values and uses an internal resampling method to determine the best value. The autoplot() method, shown in Figure 20-1, helps us understand if the default penalization method was sufficient:

```
autoplot(ens)
```

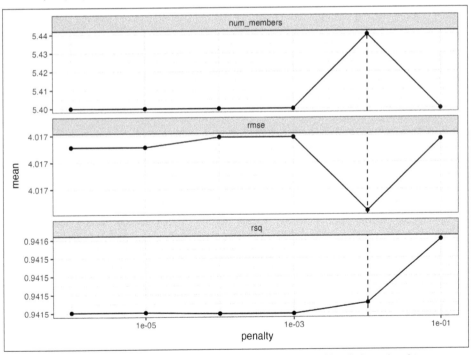

Figure 20-1. Results of using the autoplot() method on the blended stacks object.

The top panel of Figure 20-1 shows the average number of candidate ensemble members retained by the meta-learning model. We can see that the number of members is fairly constant and, as it increases, the RMSE also increases.

The default range may not have served us well here. To evaluate the meta-learning model with larger penalties, let's pass an additional option:

```
set.seed(2002)
ens <- blend_predictions(concrete_stack, penalty = 10^seq(-2, -0.5, length = 20))
```

Now, in Figure 20-2, we see a range where the ensemble model becomes worse than with our first blend (but not by much). The R^2 values increase with more members and larger penalties:

```
autoplot(ens)
```

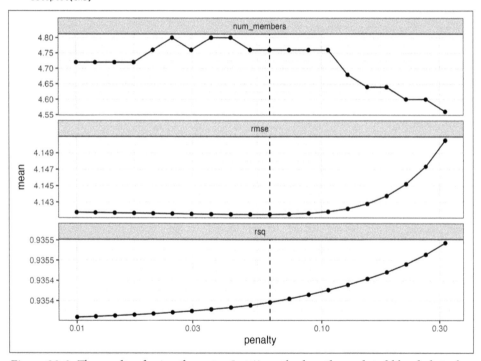

Figure 20-2. The results of using the autoplot() *method on the updated blended stacks object.*

When blending predictions using a regression model, it is common to constrain the blending parameters to be nonnegative. For these data, this constraint has the effect of eliminating many of the potential ensemble members; even at fairly low penalties, the ensemble is limited to a fraction of the original 18.

The penalty value associated with the smallest RMSE was 0.051. Printing the object shows the details of the meta-learning model:

```
ens
#> — A stacked ensemble model ————————————————
#>
#> Out of 21 possible candidate members, the ensemble retained 7.
#> Penalty: 0.0513483290743755.
#> Mixture: 1.
#>
#> The 7 highest weighted members are:
#> # A tibble: 7 × 3
#>    member                   type           weight
#>    <chr>                    <chr>           <dbl>
#> 1 boosting_1_04            boost_tree     0.727
#> 2 neural_network_1_17      mlp            0.101
#> 3 Cubist_1_25              cubist_rules   0.0906
#> 4 neural_network_1_04      mlp            0.0820
#> 5 full_quad_linear_reg_1_16 linear_reg    0.0176
#> 6 full_quad_linear_reg_1_17 linear_reg    0.00284
#> # … with 1 more row
#>
#> Members have not yet been fitted with `fit_members()`.
```

The regularized linear regression meta-learning model contained seven blending coefficients across four types of models. The `autoplot()` method can be used again to show the contributions of each model type, to produce Figure 20-3:

```
autoplot(ens, "weights") +
  geom_text(aes(x = weight + 0.01, label = model), hjust = 0) +
  theme(legend.position = "none") +
  lims(x = c(-0.01, 0.8))
```

The boosted tree and neural network models have the largest contributions to the ensemble. For this ensemble, the outcome is predicted with the equation:

$$
\begin{aligned}
\text{ensemble prediction} = {} & -0.65 \\
& + 0.77 \times \text{boost tree prediction} \\
& + 0.16 \times \text{cubist rules prediction} \\
& + 0.044 \times \text{linear reg prediction} \\
& + 0.03 \times \text{mlp prediction} \\
& + 0.013 \times \text{mars prediction}
\end{aligned}
$$

where the predictors in the equation are the predicted compressive strength values from those models.

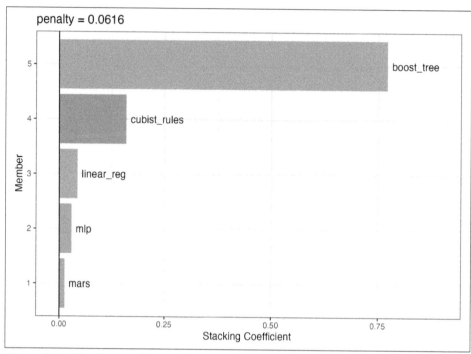

Figure 20-3. Blending coefficients for the stacking ensemble.

Fit the Member Models

The ensemble contains seven candidate members, and we now know how their predictions can be blended into a final prediction for the ensemble. However, these individual model fits have not yet been created. To be able to use the stacking model, seven additional model fits are required. These use the entire training set with the original predictors.

The seven models to be fit are:

Boosting
Number of trees = 1957, minimal node size = 8, tree depth = 7, learning rate = 0.0756, minimum loss reduction = 1.45e–07, and proportion of observations sampled = 0.679

Cubist
Number of committees = 98 and number of nearest neighbors = 2

Linear regression (quadratic features)
Amount of regularization = 6.28e–09 and proportion of lasso penalty = 0.636 (config 1)

Linear regression (quadratic features)
> Amount of regularization = 2e–09 and proportion of lasso penalty = 0.668 (config 2)

Neural network
> Number of hidden units = 14, amount of regularization = 0.0345, and number of epochs = 979 (config 1)

Neural network
> Number of hidden units = 22, amount of regularization = 2.08e–10, and number of epochs = 92 (config 2)

Neural network
> Number of hidden units = 26, amount of regularization = 0.0149, and number of epochs = 203 (config 3)

The stacks package has a function, `fit_members()`, that trains and returns these models:

```
ens <- fit_members(ens)
```

This updates the stacking object with the fitted workflow objects for each member. At this point, the stacking model can be used for prediction.

Test Set Results

Since the blending process used resampling, we can estimate that the ensemble with seven members had an estimated RMSE of 4.12. Recall from Chapter 15 that the best boosted tree had a test set RMSE of 3.33. How will the ensemble model compare on the test set? We can `predict()` to find out:

```
reg_metrics <- metric_set(rmse, rsq)
ens_test_pred <-
  predict(ens, concrete_test) %>%
  bind_cols(concrete_test)

ens_test_pred %>%
  reg_metrics(compressive_strength, .pred)
#> # A tibble: 2 × 3
#>   .metric .estimator .estimate
#>   <chr>   <chr>          <dbl>
#> # 1 rmse   standard        3.33
#> # 2 rsq    standard        0.957
```

This is moderately better than our best single model. It is fairly common for stacking to produce incremental benefits when compared to the best single model.

Chapter Summary

This chapter demonstrated how to combine different models into an ensemble for better predictive performance. The process of creating the ensemble can automatically eliminate candidate models to find a small subset that improves performance. The stacks package has a fluent interface for combining resampling and tuning results into a metamodel.

## CHAPTER 21

Inferential Analysis

 In Chapter 1, we outlined a taxonomy of models and said that most models can be categorized as descriptive, inferential, and/or predictive.

Most of the chapters in this book have focused on models from the perspective of the accuracy of predicted values, an important quality of models for all purposes but most relevant for predictive models. Inferential models are usually created not only for their predictions but also to make inferences or judgments about some component of the model, such as a coefficient value or other parameter. These results are often used to answer some (hopefully) predefined questions or hypotheses. In predictive models, predictions on holdout data are used to validate or characterize the quality of the model. Inferential methods focus on validating the probabilistic or structural assumptions that are made prior to fitting the model.

For example, in ordinary linear regression, the common assumption is that the residual values are independent and follow a Gaussian distribution with a constant variance. While you may have scientific or domain knowledge to lend credence to this assumption for your model analysis, the residuals from the fitted model are usually examined to determine if the assumption was a good idea. As a result, the methods for determining if the model's assumptions have been met are not as simple as looking at holdout predictions, although that can be very useful as well.

We will use p-values in this chapter. However, the tidymodels framework tends to promote confidence intervals over p-values as a method for quantifying the evidence for an alternative hypothesis. As previously shown in Chapter 11, Bayesian methods are often superior to both p-values and confidence intervals in terms of ease of interpretation (but they can be more computationally expensive).

There has been a push in recent years to move away from p-values in favor of other methods (Wasserstein and Lazar 2016). See Volume 73 of *The American Statistician* (*https://oreil.ly/UeukP*) for more information and discussion.

In this chapter, we describe how to use tidymodels for fitting and assessing inferential models. In some cases, the tidymodels framework can help users work with the objects produced by their models. In others, it can help assess the quality of a given model.

Inference for Count Data

To understand how tidymodels packages can be used for inferential modeling, let's focus on an example with count data. We'll use biochemistry publication data from the pscl package. These data consist of information on 915 Ph.D. biochemistry graduates and tries to explain factors that impact their academic productivity (measured via number or count of articles published within three years). The predictors include the gender of the graduate, their marital status, the number of children of the graduate that are at least 5 years old, the prestige of their department, and the number of articles produced by their mentor in the same time period. The data reflect biochemistry doctorates who finished their education between 1956 and 1963. The data are a somewhat biased sample of all of the biochemistry doctorates given during this period (based on completeness of information).

Recall that in Chapter 19 we asked the question "Is our model applicable for predicting a specific data point?" It is very important to define what populations an inferential analysis applies to. For these data, the results would likely apply to biochemistry doctorates given around the time frame that the data were collected. Does it also apply to other chemistry doctorate types (e.g., medicinal chemistry, etc.)? These are important questions to address (and document) when conducting inferential analyses.

A plot of the data shown in Figure 21-1 indicates that many graduates did not publish any articles in this time and that the outcome follows a right-skewed distribution:

```
library(tidymodels)
tidymodels_prefer()

data("bioChemists", package = "pscl")

ggplot(bioChemists, aes(x = art)) +
  geom_histogram(binwidth = 1, color = "white") +
  labs(x = "Number of articles within 3y of graduation")
```

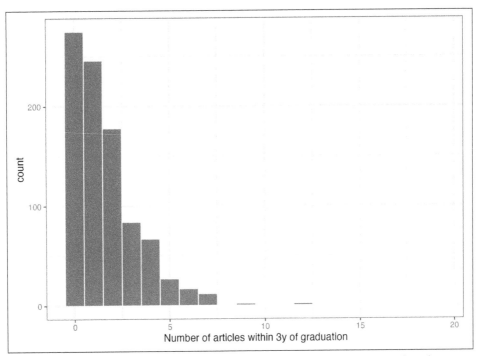

Figure 21-1. Distribution of the number of articles written within 3 years of graduation.

Since the outcome data are counts, the most common distribution assumption to make is that the outcome has a Poisson distribution. This chapter will use these data for several types of analyses.

Comparisons with Two-Sample Tests

We can start with hypothesis testing. The original author's goal with this data set on biochemistry publication data was to determine if there is a difference in publications between men and women (Long 1992). The data from the study show:

```
bioChemists %>%
  group_by(fem) %>%
  summarize(counts = sum(art), n = length(art))
#> # A tibble: 2 × 3
#>   fem    counts     n
#>   <fct>   <int> <int>
#> 1 Men       930   494
#> 2 Women     619   421
```

There were many more publications by men, although there also were more men in the data. The simplest approach to analyzing these data would be to do a two-sample comparison using the `poisson.test()` function in the stats package. It requires the counts for one or two groups.

For our application, the hypotheses to compare the two sexes are:

$$H_0 : \lambda_m = \lambda_f$$
$$H_a : \lambda_m \neq \lambda_f$$

where the λ values are the rates of publications (over the same time period).

A basic application of the test is:[1]

```
poisson.test(c(930, 619), T = 3)
#>
#>   Comparison of Poisson rates
#>
#> data:  c(930, 619) time base: 3
#> count1 = 930, expected count1 = 774, p-value = 3e-15
#> alternative hypothesis: true rate ratio is not equal to 1
#> 95 percent confidence interval:
#>   1.356 1.666
#> sample estimates:
#> rate ratio
#>      1.502
```

The function reports a p-value as well as a confidence interval for the ratio of the publication rates. The results indicate that the observed difference is greater than the experiential noise and favors H_a.

One issue with using this function is that the results come back as an `htest` object. While this type of object has a well-defined structure, it can be difficult to consume for subsequent operations such as reporting or visualizations. The most impactful tool that tidymodels offers for inferential models is the `tidy()` functions in the broom package. As previously seen, this function makes a well-formed, predictably named tibble from the object. We can `tidy()` the results of our two-sample comparison test:

```
poisson.test(c(930, 619)) %>%
  tidy()
#> # A tibble: 1 × 8
#>   estimate statistic  p.value parameter conf.low conf.high method        alternative
#>      <dbl>     <dbl>    <dbl>     <dbl>    <dbl>     <dbl> <chr>         <chr>
#> 1     1.50       930 2.73e-15      774.     1.36      1.67 Comparison o… two.sided
```

 Between the broom (*https://oreil.ly/jtbP8*) and broom.mixed (*https://oreil.ly/NIHXK*) packages, there are `tidy()` methods for more than 150 models.

---

[1] The T argument allows us to account for the time when the events (publications) were counted, which was three years for both men and women. There are more men than women in these data, but `poisson.test()` has limited functionality so more sophisticated analysis can be used to account for this difference.

While the Poisson distribution is reasonable, we might also want to assess using fewer distributional assumptions. Two methods that might be helpful are the bootstrap and permutation tests (Davison and Hinkley 1997).

The infer package, part of the tidymodels framework, is a powerful and intuitive tool for hypothesis testing (Ismay and Kim 2021). Its syntax is concise and designed for nonstatisticians.

First, we specify() that we will use the difference in the mean number of articles between the sexes and then calculate() the statistic from the data. Recall that the maximum likelihood estimator for the Poisson mean is the sample mean. The hypotheses tested here are the same as the previous test (but are conducted using a different testing procedure).

With infer, we specify the outcome and covariate, then state the statistic of interest:

```
library(infer)

observed <-
  bioChemists %>%
  specify(art ~ fem) %>%
  calculate(stat = "diff in means", order = c("Men", "Women"))
observed
#> Response: art (numeric)
#> Explanatory: fem (factor)
#> # A tibble: 1 × 1
#>     stat
#>    <dbl>
#> 1  0.412
```

From here, we compute a confidence interval for this mean by creating the bootstrap distribution via generate(); the same statistic is computed for each resampled version of the data:

```
set.seed(2101)
bootstrapped <-
  bioChemists %>%
  specify(art ~ fem)  %>%
  generate(reps = 2000, type = "bootstrap") %>%
  calculate(stat = "diff in means", order = c("Men", "Women"))
bootstrapped
#> Response: art (numeric)
#> Explanatory: fem (factor)
#> # A tibble: 2,000 × 2
#>    replicate  stat
#>        <int> <dbl>
#> 1          1 0.467
#> 2          2 0.107
#> 3          3 0.467
#> 4          4 0.308
#> 5          5 0.369
#> 6          6 0.428
#> # … with 1,994 more rows
```

A percentile interval is calculated using:

```
percentile_ci <- get_ci(bootstrapped)
percentile_ci
#> # A tibble: 1 × 2
#>   lower_ci upper_ci
#>      <dbl>    <dbl>
#> 1    0.158    0.653
```

The infer package has a high-level API for showing the analysis results, as shown in Figure 21-2:

```
visualize(bootstrapped) +
    shade_confidence_interval(endpoints = percentile_ci)
```

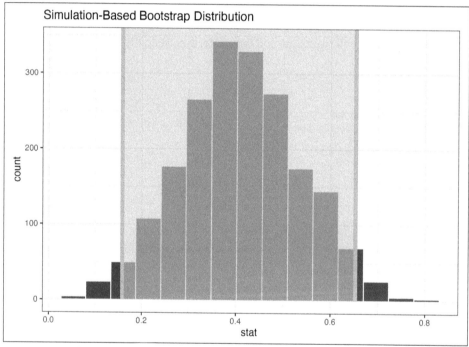

Figure 21-2. The bootstrap distribution of the difference in means. The highlighted region is the confidence interval.

Since the interval visualized in in Figure 21-2 does not include zero, these results indicate that men have published more articles than women.

If we require a p-value, the infer package can compute the value via a permutation test, shown in the following code. The syntax is very similar to the bootstrapping code we used earlier. We add a `hypothesize()` verb to state the type of assumption to test, and the `generate()` call contains an option to shuffle the data:

```
set.seed(2102)
permuted <-
```

```
bioChemists %>%
  specify(art ~ fem)  %>%
  hypothesize(null = "independence") %>%
  generate(reps = 2000, type = "permute") %>%
  calculate(stat = "diff in means", order = c("Men", "Women"))
permuted
#> Response: art (numeric)
#> Explanatory: fem (factor)
#> Null Hypothesis: independence
#> # A tibble: 2,000 × 2
#>   replicate       stat
#>       <int>      <dbl>
#> 1         1    0.201
#> 2         2   -0.133
#> 3         3    0.109
#> 4         4   -0.195
#> 5         5   -0.00128
#> 6         6   -0.102
#> # … with 1,994 more rows
```

The following visualization code is also very similar to the bootstrap approach. This code generates Figure 21-3 where the vertical line signifies the observed value:

```
visualize(permuted) +
    shade_p_value(obs_stat = observed, direction = "two-sided")
```

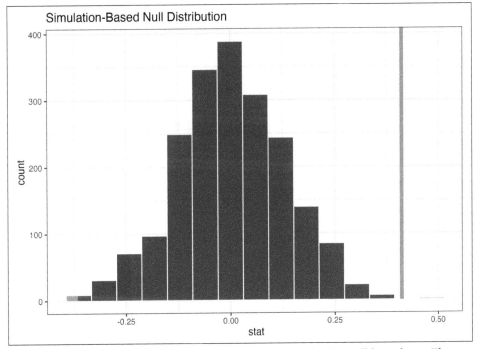

Figure 21-3. Empirical distribution of the test statistic under the null hypothesis. The vertical line indicates the observed test statistic.

The actual p-value is:

```
permuted %>%
  get_p_value(obs_stat = observed, direction = "two-sided")
#> # A tibble: 1 × 1
#>   p_value
#>     <dbl>
#> 1   0.002
```

The vertical line representing the null hypothesis in Figure 21-3 is far away from the permutation distribution. This means, if in fact the null hypothesis were true, the likelihood is exceedingly small of observing data at least as extreme as what is at hand.

The two-sample tests shown in this section are probably suboptimal because they do not account for other factors that might explain the observed relationship between publication rate and sex. Let's move to a more complex model that can consider additional covariates.

Log-Linear Models

The focus of the rest of this chapter will be on a generalized linear model (Dobson 1999) where we assume the counts follow a Poisson distribution. For this model, the covariates/predictors enter the model in a log-linear fashion:

$$\log(\lambda) = \beta_0 + \beta_1 x_1 + \ldots + \beta_p x_p$$

where λ is the expected value of the counts.

Let's fit a simple model that contains all of the predictor columns. The poisson-reg package, a parsnip extension package in tidymodels, will create this model specification:

```
library(poissonreg)

# default engine is 'glm'
log_lin_spec <- poisson_reg()

log_lin_fit <-
  log_lin_spec %>%
  fit(art ~ ., data = bioChemists)
log_lin_fit
#> parsnip model object
#>
#>
#> Call:  stats::glm(formula = art ~ ., family = stats::poisson, data = data)
#>
#> Coefficients:
#> (Intercept)     femWomen    marMarried         kid5          phd         ment
#>      0.3046      -0.2246        0.1552      -0.1849       0.0128       0.0255
#>
#> Degrees of Freedom: 914 Total (i.e., Null);  909 Residual
```

```
#> Null Deviance:      1820
#> Residual Deviance: 1630  AIC: 3310
```

The `tidy()` method succinctly summarizes the coefficients for the model (along with 90% confidence intervals):

```
tidy(log_lin_fit, conf.int = TRUE, conf.level = 0.90)
#> # A tibble: 6 × 7
#>   term          estimate std.error statistic  p.value conf.low conf.high
#>   <chr>            <dbl>     <dbl>     <dbl>    <dbl>    <dbl>     <dbl>
#> 1 (Intercept)    0.305    0.103      2.96   3.10e- 3   0.134     0.473
#> 2 femWomen      -0.225    0.0546    -4.11   3.92e- 5  -0.315    -0.135
#> 3 marMarried     0.155    0.0614     2.53   1.14e- 2   0.0545    0.256
#> 4 kid5          -0.185    0.0401    -4.61   4.08e- 6  -0.251    -0.119
#> 5 phd            0.0128   0.0264     0.486  6.27e- 1  -0.0305    0.0563
#> 6 ment           0.0255   0.00201   12.7    3.89e-37   0.0222    0.0288
```

In the previous output, the p-values correspond to separate hypothesis tests for each parameter:

$$H_0 : \beta_j = 0$$
$$H_a : \beta_j \neq 0$$

for each of the model parameters. Looking at these results, phd (the prestige of their department) may not have any relationship with the outcome.

While the Poisson distribution is the routine assumption for data like these, it may be beneficial to conduct a rough check of the model assumptions by fitting the models without using the Poisson likelihood to calculate the confidence intervals. The rsample package has a convenience function to compute bootstrap confidence intervals for `lm()` and `glm()` models. We can use this function, while explicitly declaring `family = poisson`, to compute a large number of model fits. By default, we compute a 90% confidence bootstrap-t interval (percentile intervals are also available):

```
set.seed(2103)
glm_boot <-
  reg_intervals(art ~ ., data = bioChemists, model_fn = "glm", family = poisson)
glm_boot
#> # A tibble: 5 × 6
#>   term         .lower .estimate  .upper .alpha .method
#>   <chr>         <dbl>     <dbl>   <dbl>  <dbl> <chr>
#> 1 femWomen    -0.358    -0.226  -0.0856  0.05 student-t
#> 2 kid5        -0.298    -0.184  -0.0789  0.05 student-t
#> 3 marMarried   0.000264  0.155   0.317   0.05 student-t
#> 4 ment         0.0182    0.0256  0.0322  0.05 student-t
#> 5 phd         -0.0707    0.0130  0.102   0.05 student-t
```

When we compare these results (in Figure 21-4) to the purely parametric results from `glm()`, the bootstrap intervals are somewhat wider. If the data were truly Poisson, these intervals would have more similar widths.

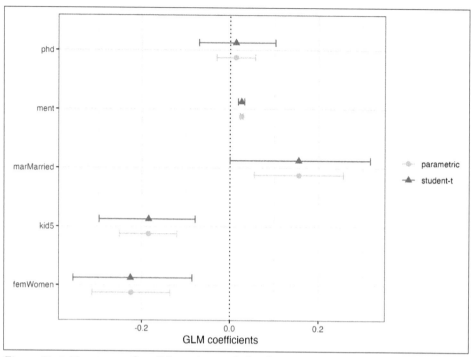

Figure 21-4. Two types of confidence intervals for the Poisson regression model.

Determining which predictors to include in the model is a difficult problem. One approach is to conduct likelihood ratio tests (LRT) (McCullagh and Nelder 1989) between nested models. Based on the confidence intervals, we have evidence that a simpler model without phd may be sufficient. Let's fit a smaller model, then conduct a statistical test:

$$H_0 : \beta_{phd} = 0$$
$$H_a : \beta_{phd} \neq 0$$

This hypothesis was previously tested when we showed the tidied results for `log_lin_fit`. That particular approach used results from a single model fit via a Wald statistic (i.e., the parameter divided by its standard error). For that approach, the p-value was 0.63. We can tidy the results for the LRT to get the p-value:

```
log_lin_reduced <-
  log_lin_spec %>%
  fit(art ~ ment + kid5 + fem + mar, data = bioChemists)

anova(
  extract_fit_engine(log_lin_reduced),
  extract_fit_engine(log_lin_fit),
  test = "LRT"
```

```
) %>%
  tidy()
#> # A tibble: 2 × 5
#>   Resid..Df Resid..Dev    df Deviance p.value
#>       <dbl>      <dbl> <dbl>    <dbl>   <dbl>
#> 1       910      1635.    NA       NA      NA
#> 2       909      1634.     1    0.236   0.627
```

The results are the same and, based on these and the confidence interval for this parameter, we'll exclude phd from further analyses since it does not appear to be associated with the outcome.

A More Complex Model

We can move into even more complex models within our tidymodels approach. For count data, there are occasions where the number of zero counts is larger than what a simple Poisson distribution would prescribe. A more complex model appropriate for this situation is the zero-inflated Poisson (ZIP) model; see Mullahy (1986), Lambert (1992), and Zeileis, Kleiber, and Jackman (2008). Here, there are two sets of covariates: one for the count data and others that affect the probability (denoted as π) of zeros. The equation for the mean λ is:

$$\lambda = 0\pi + (1 - \pi)\lambda_{nz}$$

where

$$\log(\lambda_{nz}) = \beta_0 + \beta_1 x_1 + \dots + \beta_p x_p$$
$$\log\left(\frac{\pi}{1-\pi}\right) = \gamma_0 + \gamma_1 z_1 + \dots + \gamma_q z_q$$

and the x covariates affect the count values while the z covariates influence the probability of a zero. The two sets of predictors do not need to be mutually exclusive.

We'll fit a model with a full set of z covariates:

```
zero_inflated_spec <- poisson_reg() %>% set_engine("zeroinfl")

zero_inflated_fit <-
  zero_inflated_spec %>%
  fit(art ~ fem + mar + kid5 + ment | fem + mar + kid5 + phd + ment,
      data = bioChemists)

zero_inflated_fit
#> parsnip model object
#>
#>
#> Call:
#> pscl::zeroinfl(formula = art ~ fem + mar + kid5 + ment | fem + mar + kid5 +
#>     phd + ment, data = data)
```

```
#>
#> Count model coefficients (poisson with log link):
#> (Intercept)     femWomen    marMarried         kid5        ment
#>       0.621      -0.209         0.105       -0.143       0.018
#>
#> Zero-inflation model coefficients (binomial with logit link):
#> (Intercept)     femWomen    marMarried         kid5         phd         ment
#>     -0.6086       0.1093       -0.3529       0.2195      0.0124      -0.1351
```

Since the coefficients for this model are also estimated using maximum likelihood, let's try to use another likelihood ratio test to understand if the new model terms are helpful. We will *simultaneously* test that:

$$H_0 : \gamma_1 = 0, \gamma_2 = 0, \cdots, \gamma_5 = 0$$
$$H_a : \text{at least one } \gamma \neq 0$$

Let's try ANOVA again:

```
anova(
  extract_fit_engine(zero_inflated_fit),
  extract_fit_engine(log_lin_reduced),
  test = "LRT"
) %>%
  tidy()
#> Error in UseMethod("anova"): no applicable method for 'anova' applied to an
    object of class "zeroinfl"
```

An anova() method isn't implemented for zeroinfl objects!

An alternative is to use an *information criterion statistic*, such as the Akaike information criterion (AIC) (Claeskens 2016). This computes the log-likelihood (from the training set) and penalizes that value based on the training set size and the number of model parameters. In R's parameterization, smaller AIC values are better. In this case, we are not conducting a formal statistical test but *estimating* the ability of the data to fit the data.

The results indicate that the ZIP model is preferable:

```
zero_inflated_fit %>% extract_fit_engine() %>% AIC()
#> [1] 3232
log_lin_reduced    %>% extract_fit_engine() %>% AIC()
#> [1] 3312
```

However, it's hard to contextualize this pair of single values and assess *how* different they actually are. To solve this problem, we'll resample a large number of each of these two models. From these, we can compute the AIC values for each and determine how often the results favor the ZIP model. Basically, we will be characterizing the uncertainty of the AIC statistics to gauge their difference relative to the noise in the data.

We'll also compute more bootstrap confidence intervals for the parameters in a bit so we specify the `apparent = TRUE` option when creating the bootstrap samples. This is required for some types of intervals.

First, we create the 4,000 model fits:

```
zip_form <- art ~ fem + mar + kid5 + ment | fem + mar + kid5 + phd + ment
glm_form <- art ~ fem + mar + kid5 + ment

set.seed(2104)
bootstrap_models <-
  bootstraps(bioChemists, times = 2000, apparent = TRUE) %>%
  mutate(
    glm = map(splits, ~ fit(log_lin_spec,      glm_form, data = analysis(.x))),
    zip = map(splits, ~ fit(zero_inflated_spec, zip_form, data = analysis(.x)))
  )
bootstrap_models
#> # Bootstrap sampling with apparent sample
#> # A tibble: 2,001 × 4
#>   splits            id            glm      zip
#>   <list>            <chr>         <list>   <list>
#> 1 <split [915/355]> Bootstrap0001 <fit[+]> <fit[+]>
#> 2 <split [915/333]> Bootstrap0002 <fit[+]> <fit[+]>
#> 3 <split [915/337]> Bootstrap0003 <fit[+]> <fit[+]>
#> 4 <split [915/344]> Bootstrap0004 <fit[+]> <fit[+]>
#> 5 <split [915/351]> Bootstrap0005 <fit[+]> <fit[+]>
#> 6 <split [915/354]> Bootstrap0006 <fit[+]> <fit[+]>
#> # … with 1,995 more rows
```

Now we can extract the model fits and their corresponding AIC values:

```
bootstrap_models <-
  bootstrap_models %>%
  mutate(
    glm_aic = map_dbl(glm, ~ extract_fit_engine(.x) %>% AIC()),
    zip_aic = map_dbl(zip, ~ extract_fit_engine(.x) %>% AIC())
  )
mean(bootstrap_models$zip_aic < bootstrap_models$glm_aic)
#> [1] 1
```

It seems definitive from these results that accounting for the excessive number of zero counts is a good idea.

We could have used `fit_resamples()` or a workflow set to conduct these computations. In this section, we used `mutate()` and `map()` to compute the models to demonstrate how one might use tidymodels tools for models that are not supported by one of the parsnip packages.

Since we have computed the resampled model fits, let's create bootstrap intervals for the zero probability model coefficients (i.e., the γ_j). We can extract these with the `tidy()` method and use the `type = "zero"` option to obtain these estimates:

```
bootstrap_models <-
  bootstrap_models %>%
  mutate(zero_coefs = map(zip, ~ tidy(.x, type = "zero")))

# One example:
bootstrap_models$zero_coefs[[1]]
#> # A tibble: 6 × 6
#>   term         type  estimate std.error statistic  p.value
#>   <chr>        <chr>    <dbl>     <dbl>     <dbl>    <dbl>
#> 1 (Intercept) zero   -0.128     0.497    -0.257 0.797
#> 2 femWomen     zero   -0.0764    0.319    -0.240 0.811
#> 3 marMarried   zero   -0.112     0.365    -0.307 0.759
#> 4 kid5         zero    0.270     0.186     1.45  0.147
#> 5 phd          zero   -0.178     0.132    -1.35  0.177
#> 6 ment         zero   -0.123     0.0315   -3.91  0.0000935
```

It's a good idea to visualize the bootstrap distributions of the coefficients, as in Figure 21-5:

```
bootstrap_models %>%
  unnest(zero_coefs) %>%
  ggplot(aes(x = estimate)) +
  geom_histogram(bins = 25, color = "white") +
  facet_wrap(~ term, scales = "free_x") +
  geom_vline(xintercept = 0, lty = 2, color = "gray70")
```

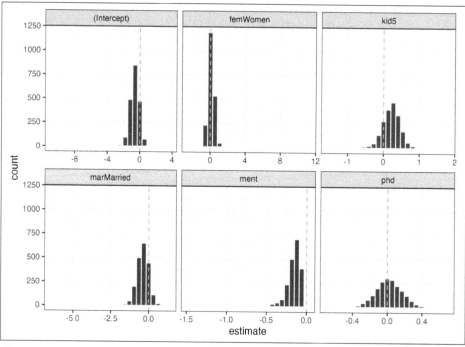

Figure 21-5. Bootstrap distributions of the ZIP model coefficients. The vertical lines indicate the observed estimates.

One of the covariates (ment) that appears to be important has a very skewed distribution. The extra space in some of the facets indicates there are some outliers in the estimates. This *might* occur when models did not converge; those results probably should be excluded from the resamples. For the results visualized in Figure 21-5, the outliers are due only to extreme parameter estimates; all of the models converged.

The rsample package contains a set of functions named int_*() that compute different types of bootstrap intervals. Since the tidy() method contains standard error estimates, the bootstrap-t intervals can be computed. We'll also compute the standard percentile intervals. By default, 90% confidence intervals are computed:

```
bootstrap_models %>% int_pctl(zero_coefs)
#> # A tibble: 6 × 6
#>   term          .lower .estimate  .upper .alpha .method
#>   <chr>          <dbl>     <dbl>   <dbl>  <dbl> <chr>
#> 1 (Intercept)   -1.75    -0.621   0.423   0.05 percentile
#> 2 femWomen      -0.521    0.115   0.818   0.05 percentile
#> 3 kid5          -0.327    0.218   0.677   0.05 percentile
#> 4 marMarried    -1.20    -0.381   0.362   0.05 percentile
#> 5 ment          -0.401   -0.162  -0.0513  0.05 percentile
#> 6 phd           -0.276    0.0220  0.327   0.05 percentile
bootstrap_models %>% int_t(zero_coefs)
#> # A tibble: 6 × 6
#>   term          .lower .estimate  .upper .alpha .method
#>   <chr>          <dbl>     <dbl>   <dbl>  <dbl> <chr>
#> 1 (Intercept)   -1.61    -0.621   0.321   0.05 student-t
#> 2 femWomen      -0.482    0.115   0.671   0.05 student-t
#> 3 kid5          -0.211    0.218   0.599   0.05 student-t
#> 4 marMarried    -0.988   -0.381   0.290   0.05 student-t
#> 5 ment          -0.324   -0.162  -0.0275  0.05 student-t
#> 6 phd           -0.274    0.0220  0.291   0.05 student-t
```

From these results, we can get a good idea of which predictor(s) to include in the zero count probability model. It may be sensible to refit a smaller model to assess if the bootstrap distribution for ment is still skewed.

More Inferential Analysis

This chapter demonstrated just a small subset of what is available for inferential analysis in tidymodels and has focused on resampling and frequentist methods. Arguably, Bayesian analysis is a very effective and often superior approach for inference. A variety of Bayesian models are available via parsnip. Additionally, the multilevelmod package enables users to fit hierarchical Bayesian and non-Bayesian models (e.g., mixed models). The broom.mixed and tidybayes packages are excellent tools for extracting data for plots and summaries. Finally, for data sets with a single hierarchy, such as simple longitudinal or repeated measures data, rsample's group_vfold_cv() function facilitates straightforward out-of-sample characterizations of model performance.

Chapter Summary

The tidymodels framework is for more than predictive modeling alone. Packages and functions from tidymodels can be used for hypothesis testing, as well as fitting and assessing inferential models. The tidymodels framework provides support for working with non-tidymodels R models, and can help assess the statistical qualities of your models.

Recommended Preprocessing

The type of preprocessing needed depends on the type of model being fit. For example, models that use distance functions or dot products should have all of their predictors on the same scale so that distance is measured appropriately.

You can learn more about each of these models, and others that might be available, at the tinymodels website (*https://oreil.ly/Eco3u*).

This Appendix provides recommendations for baseline levels of preprocessing that are needed for various model functions. In Table A-1, the preprocessing methods are categorized as:

Dummy
Do qualitative predictors require a numeric encoding (e.g., via dummy variables or other methods)?

ZV
Should columns with a single unique (i.e., zero variance) value be removed?

Impute
If some predictors are missing, should they be estimated via imputation?

Decorrelate
If there are correlated predictors, should this correlation be mitigated? This might mean filtering out predictors, using principal component analysis, or a model-based technique (e.g., regularization).

Normalize
Should predictors be centered and scaled?

Transform
Is it helpful to transform predictors to be more symmetric?

The information in Table A-1 is not exhaustive and somewhat depends on the implementation. For example, as noted in the table's footnotes, some models may not require a particular preprocessing operation but the implementation may require it. In the table, ✓ indicates that the method is required for the model and × indicates that it is not. The ○ symbol means that the model *may* be helped by the technique but it is not required.

Table A-1. Preprocessing methods for different models

Model	Dummy	ZV	Impute	Decorrelate	Normalize	Transform
bag_mars()	✓	×	✓	○	×	○
bag_tree()	×	×	×	○[a]	×	×
bart()	×	×	×	○[a]	×	×
boost_tree()	×[b]	○	✓[b]	○[a]	×	×
C5_rules()	×	×	×	×	×	×
cubist_rules()	×	×	×	×	×	×
decision_tree()	×	×	×	○[a]	×	×
discrim_flexible()	✓	×	✓	✓	×	○
discrim_linear()	✓	✓	✓	✓	×	○
discrim_regularized()	✓	✓	✓	✓	×	○
gen_additive_mod()	✓	✓	✓	✓	×	○
linear_reg()	✓	✓	✓	✓	×	○
logistic_reg()	✓	✓	✓	✓	×	○
mars()	✓	×	✓	○	×	○
mlp()	✓	✓	✓	✓	✓	✓
multinom_reg()	✓	✓	✓	✓	×[b]	○
naive_Bayes()	×	✓	✓	○[a]	×	×
nearest_neighbor()	✓	✓	✓	○	✓	✓
pls()	✓	✓	✓	×	✓	✓
poisson_reg()	✓	✓	✓	✓	×	○
rand_forest()	×	○	✓[b]	○[a]	×	×
rule_fit()	✓	×	✓	○[a]	✓	×
svm_*()	✓	✓	✓	✓	✓	✓

[a] Decorrelating predictors may not help improve performance. However, fewer correlated predictors can improve the estimation of variable importance scores; see Fig. 11.4 of Kuhn and Johnson (2020) (*https://oreil.ly/PDlm1*). Essentially, the selection of highly correlated predictors is almost random.

[b] The needed preprocessing for these models depends on the implementation. Specifically: *theoretically*, any tree-based model does not require imputation. However, many tree ensemble implementations require imputation. While tree-based boosting methods generally do not require the creation of dummy variables, models using the xgboost engine do.

References

Aboumatar, H., and R. Wise. 2019. "Notice of Retraction. Aboumatar et al. Effect of a Program Combining Transitional Care and Long-Term Self-Management Support on Outcomes of Hospitalized Patients with Chronic Obstructive Pulmonary Disease: A Randomized Clinical Trial." *JAMA* 322 (14): 1417–18.

Abrams, B. 2003. "The Pit of Success" (*https://oreil.ly/yGYrb*).

Baggerly, K., and K. Coombes. 2009. "Deriving Chemosensitivity from Cell Lines: Forensic Bioinformatics and Reproducible Research in High-Throughput Biology." *The Annals of Applied Statistics* 3 (4): 1309–34.

Bartley, M., E. Hanks, E. Schliep, P. Sorrano, and T. Wagner. 2019. "Identifying and Characterizing Extrapolation in Multivariate Response Data" (*https://oreil.ly/ZPtpZ*). *PLOS ONE* 14 (12): 1–20.

Biecek, P., and T. Burzykowski. 2021. *Explanatory Model Analysis* (*https://ema.drwhy.ai*). New York: Chapman; Hall/CRC.

Bohachevsky, I., M. Johnson, and M. Stein. 1986. "Generalized Simulated Annealing for Function Optimization." *Technometrics* 28 (3): 209–17.

Bolstad, B. 2004. *Low-Level Analysis of High-Density Oligonucleotide Array Data: Background, Normalization and Summarization*. Berkley: University of California Press.

Box, George E. P., W. Hunter, and J. Hunter. 2005. *Statistics for Experimenters: An Introduction to Design, Data Analysis, and Model Building*. New York: Wiley.

Bradley, R., and M. Terry. 1952. "Rank Analysis of Incomplete Block Designs: I. The Method of Paired Comparisons." *Biometrika* 39 (3/4): 324–45.

Breiman, L. 1996a. "Bagging Predictors." *Machine Learning* 24 (2): 123–40.

———. 1996b. "Stacked Regressions." *Machine Learning* 24 (1): 49–64.

———. 2001a. "Random Forests." *Machine Learning* 45 (1): 5–32.

———. 2001b. "Statistical Modeling: The Two Cultures." *Statistical Science* 16 (3): 199–231.

Carlson, B. 2012. "Putting Oncology Patients at Risk." *Biotechnology Healthcare* 9 (3): 17–21.

Chambers, J. 1998. *Programming with Data: A Guide to the S Language*. Berlin, Heidelberg: Springer-Verlag.

Chambers, J., and T. Hastie, eds. 1992. *Statistical Models in S*. Boca Raton, FL: CRC Press.

Claeskens, G. 2016. "Statistical Model Choice." *Annual Review of Statistics and Its Application* 3: 233–56.

Cleveland, W. 1979. "Robust Locally Weighted Regression and Smoothing Scatterplots." *Journal of the American Statistical Association* 74 (368): 829–36.

Craig-Schapiro, R., M. Kuhn, C. Xiong, E. Pickering, J. Liu, T. Misko, R. Perrin, et al. 2011. "Multiplexed Immunoassay Panel Identifies Novel CSF Biomarkers for Alzheimer's Disease Diagnosis and Prognosis." *PLoS ONE* 6 (4): e18850.

Cybenko, G. 1989. "Approximation by Superpositions of a Sigmoidal Function." *Mathematics of Control, Signals and Systems* 2 (4): 303–14.

Danowski, T., J. Aarons, J. Hydovitz, and J. Wingert. 1970. "Utility of Equivocal Glucose Tolerances." *Diabetes* 19 (7): 524–26.

Davison, A., and D. Hinkley. 1997. *Bootstrap Methods and Their Application*. Vol. 1. Cambridge: Cambridge University Press.

De Cock, D. 2011. "Ames, Iowa: Alternative to the Boston Housing Data as an End of Semester Regression Project." *Journal of Statistics Education* 19 (3).

Dobson, A. 1999. *An Introduction to Generalized Linear Models*. Boca Raton, FL: Chapman; Hall.

Durrleman, S., and R. Simon. 1989. "Flexible Regression Models with Cubic Splines." *Statistics in Medicine* 8 (5): 551–61.

Faraway, J. 2016. *Extending the Linear Model with R: Generalized Linear, Mixed Effects and Nonparametric Regression Models*. Boca Raton, FL: CRC Press.

Fox, J. 2008. *Applied Regression Analysis and Generalized Linear Models*. 2nd ed. Thousand Oaks, CA: Sage.

Frazier, R. 2018. "A Tutorial on Bayesian Optimization" (*https://oreil.ly/BIzLi*).

Freund, Y., and R. Schapire. 1997. "A Decision-Theoretic Generalization of On-Line Learning and an Application to Boosting." *Journal of Computer and System Sciences* 55 (1): 119–39.

Friedman, J. 1991. "Multivariate Adaptive Regression Splines." *The Annals of Statistics* 19 (1): 1–141.

———. 2001. "Greedy Function Approximation: A Gradient Boosting Machine." *Annals of Statistics* 29 (5): 1189–1232.

Friedman, J., T. Hastie, and R. Tibshirani. 2010. "Regularization Paths for Generalized Linear Models via Coordinate Descent." *Journal of Statistical Software* 33 (1): 1–22.

Geladi, P., and B. Kowalski. 1986. "Partial Least-Squares Regression: A Tutorial." *Analytica Chimica Acta* 185: 1–17.

Gentleman, R., V. Carey, W. Huber, R. Irizarry, and S. Dudoit. 2005. *Bioinformatics and Computational Biology Solutions Using R and Bioconductor*. Berlin, Heidelberg: Springer.

Good, I. J. 1985. "Weight of Evidence: A Brief Survey." *Bayesian Statistics* 2: 249–70.

Goodfellow, I., Y. Bengio, and A. Courville. 2016. *Deep Learning*. Cambridge, MA: MIT Press.

Guo, C., and F. Berkhahn. 2016. "Entity Embeddings of Categorical Variables" (*https://oreil.ly/WcSdC*).

Hand, D., and R. Till. 2001. "A Simple Generalisation of the Area Under the ROC Curve for Multiple Class Classification Problems." *Machine Learning* 45 (August): 171–86.

Hill, A., P. LaPan, Y. Li, and S. Haney. 2007. "Impact of Image Segmentation on High-Content Screening Data Quality for SK-BR-3 Cells." *BMC Bioinformatics* 8 (1): 340.

Ho, T. 1995. "Random Decision Forests." In *Proceedings of 3rd International Conference on Document Analysis and Recognition*, 1:278–82. IEEE.

Hosmer, D., and S. Lemeshow. 2000. *Applied Logistic Regression*. New York: John Wiley and Sons.

Hvitfeldt, E., and J. Silge. 2021. *Supervised Machine Learning for Text Analysis in R* (*https://smltar.com*). A Chapman & Hall Book/CRC Press/Creative Commons.

Hyndman, R., and G. Athanasopoulos. 2018. *Forecasting: Principles and Practice*. Melbourne: OTexts.

Ismay, C., and A. Kim. 2021. *Statistical Inference via Data Science: A ModernDive into R and the Tidyverse (https://moderndive.com)*. Chapman & Hall/CRC Press/Creative Commons.

Jaworska, J., N. Nikolova-Jeliazkova, and T. Aldenberg. 2005. "QSAR Applicability Domain Estimation by Projection of the Training Set in Descriptor Space: A Review." *Alternatives to Laboratory Animals* 33 (5): 445–59.

Johnson, D., P. Eckart, N. Alsamadisi, H. Noble, C. Martin, and R. Spicer. 2018. "Polar Auxin Transport Is Implicated in Vessel Differentiation and Spatial Patterning During Secondary Growth in Populus." *American Journal of Botany* 105 (2): 186–96.

Joseph, V., E. Gul, and S. Ba. 2015. "Maximum Projection Designs for Computer Experiments." *Biometrika* 102 (2): 371–80.

Jungsu, K., D. Basak, and D. Holtzman. 2009. "The Role of Apolipoprotein E in Alzheimer's Disease." *Neuron* 63 (3): 287–303.

Kerleguer, A., J.-L. Koeck, M. Fabre, P. Gérôme, R. Teyssou, and V. Hervé. 2003. "Use of Equivocal Zone in Interpretation of Results of the Amplified Mycobacterium Tuberculosis Direct Test for Diagnosis of Tuberculosis." *Journal of Clinical Microbiology* 41 (4): 1783–84.

Kirkpatrick, S., D. Gelatt, and M. Vecchi. 1983. "Optimization by Simulated Annealing." *Science* 220 (4598): 671–80.

Koklu, M., and I. A. Ozkan. 2020. "Multiclass Classification of Dry Beans Using Computer Vision and Machine Learning Techniques." *Computers and Electronics in Agriculture* 174: 105507.

Krueger, T., D. Panknin, and M. Braun. 2015. "Fast Cross-Validation via Sequential Testing." *Journal of Machine Learning Research* 16 (33): 1103–55.

Kruschke, J. 2014. *Doing Bayesian Data Analysis: A Tutorial with R, JAGS, and Stan.* Cambridge, MA: Academic Press.

Kruschke, J., and T. Liddell. 2018. "The Bayesian New Statistics: Hypothesis Testing, Estimation, Meta-Analysis, and Power Analysis from a Bayesian Perspective." *Psychonomic Bulletin and Review* 25 (1): 178–206.

Kuhn, Max. 2014. "Futility Analysis in the Cross-Validation of Machine Learning Models" (*https://oreil.ly/J3NbV*).

Kuhn, M., and K. Johnson. 2013. *Applied Predictive Modeling.* New York: Springer.

———. 2020. *Feature Engineering and Selection: A Practical Approach for Predictive Models.* Boca Raton, FL: CRC Press.

Lambert, D. 1992. "Zero-Inflated Poisson Regression, with an Application to Defects in Manufacturing." *Technometrics* 34 (1): 1–14.

Littell, R., J. Pendergast, and R. Natarajan. 2000. "Modelling Covariance Structure in the Analysis of Repeated Measures Data." *Statistics in Medicine* 19 (13): 1793–1819.

Long, J. 1992. "Measures of Sex Differences in Scientific Productivity." *Social Forces* 71 (1): 159–78.

Lundberg, S. M., and S.-I. Lee. 2017. "A Unified Approach to Interpreting Model Predictions." In *Proceedings of the 31st International Conference on Neural Information Processing Systems*, 4768–77. NIPS'17. Red Hook, NY: Curran Associates Inc.

Mangiafico, S. 2015. "An R Companion for the Handbook of Biological Statistics" (*https://oreil.ly/H9Hny*).

Maron, O., and A. Moore. 1994. "Hoeffding Races: Accelerating Model Selection Search for Classification and Function Approximation." In *Advances in Neural Information Processing Systems*, 59–66, Morgan Kaufmann Publishers.

McCullagh, P., and J. Nelder. 1989. *Generalized Linear Models*. London: Chapman & Hall.

McDonald, J. 2009. *Handbook of Biological Statistics*. Baltimore, MD: Sparky House Publishing.

McElreath, R. 2020. *Statistical Rethinking: A Bayesian Course with Examples in R and Stan*. Boca Raton, FL: CRC Press.

McInnes, L., J. Healy, and J. Melville. 2020. "UMAP: Uniform Manifold Approximation and Projection for Dimension Reduction" (*https://oreil.ly/LQxrn*).

McKay, M., R. Beckman, and W. Conover. 1979. "A Comparison of Three Methods for Selecting Values of Input Variables in the Analysis of Output from a Computer Code." *Technometrics* 21 (2): 239–45.

Micci-Barreca, D. 2001. "A Preprocessing Scheme for High-Cardinality Categorical Attributes in Classification and Prediction Problems" (*https://oreil.ly/kiZ3t*). *SIGKDD Explor. Newsl.* 3 (1): 27–32.

Molnar, Christopher. 2020. *Interpretable Machine Learning* (*https://oreil.ly/v6SMW*). Independently published.

Mullahy, J. 1986. "Specification and Testing of Some Modified Count Data Models." *Journal of Econometrics* 33 (3): 341–65.

Netzeva, T., A. Worth, T. Aldenberg, R. Benigni, M. Cronin, P. Gramatica, J. Jaworska, et al. 2005. "Current Status of Methods for Defining the Applicability Domain of (Quantitative) Structure-Activity Relationships: The Report and Recommendations of ECVAM Workshop 52." *Alternatives to Laboratory Animals* 33 (2): 155–73.

Olsson, D., and L. Nelson. 1975. "The Nelder–Mead Simplex Procedure for Function Minimization." *Technometrics* 17 (1): 45–51.

Opitz, J., and S. Burst. 2019. "Macro F1 and Macro F1" (*https://oreil.ly/dSGPZ*).

R Core Team. 2014. *R: A Language and Environment for Statistical Computing* (*http://www.R-project.org*). Vienna, Austria: R Foundation for Statistical Computing.

Rasmussen, C., and C. Williams. 2006. *Gaussian Processes for Machine Learning.* Cambridge, MA: MIT Press.

Santner, T., B. Williams, W. Notz, and B. Williams. 2003. *The Design and Analysis of Computer Experiments.* New York: Springer.

Schmidberger, M., M. Morgan, D. Eddelbuettel, H. Yu, L. Tierney, and U. Mansmann. 2009. "State of the Art in Parallel Computing with R" (*https://oreil.ly/PjhJ6*). *Journal of Statistical Software* 31 (1): 1–27.

Schulz, E., M. Speekenbrink, and A. Krause. 2018. "A Tutorial on Gaussian Process Regression: Modelling, Exploring, and Exploiting Functions." *Journal of Mathematical Psychology* 85: 1–16.

Shahriari, B., K. Swersky, Z. Wang, R. P. Adams, and N. de Freitas. 2016. "Taking the Human Out of the Loop: A Review of Bayesian Optimization." *Proceedings of the IEEE* 104 (1): 148–75.

Shewry, M., and H. Wynn. 1987. "Maximum Entropy Sampling." *Journal of Applied Statistics* 14 (2): 165–70.

Shmueli, G. 2010. "To Explain or to Predict?" *Statistical Science* 25(3): 289–310.

Symons, S., and R. G. Fulcher. 1988. "Determination of Wheat Kernel Morphological Variation by Digital Image Analysis: I. Variation in Eastern Canadian Milling Quality Wheats." *Journal of Cereal Science* 8 (3): 211–18.

Thomas, R., and D. Uminsky. 2020. "The Problem with Metrics Is a Fundamental Problem for AI" (*https://oreil.ly/oyoJa*).

Tibshirani, Robert. 1996. "Regression Shrinkage and Selection via the Lasso" (*https://oreil.ly/1uBxF*). *Journal of the Royal Statistical Society. Series B (Methodological)* 58 (1): 267–88.

Van Laarhoven, P., and E. Aarts. 1987. "Simulated Annealing." *Simulated Annealing: Theory and Applications*, edited by R. Chianti, 7–15. Springer.

Wasserstein, R., and N. Lazar. 2016. "The ASA Statement on p-Values: Context, Process, and Purpose." *The American Statistician* 70 (2): 129–33.

Weinberger, K., A. Dasgupta, J. Langford, A. Smola, and J. Attenberg. 2009. "Feature Hashing for Large Scale Multitask Learning." In *Proceedings of the 26th Annual International Conference on Machine Learning*, 1113–20. ACM.

Wickham, H. 2019. *Advanced R (https://oreil.ly/ZImoS)*. 2nd ed. New York: Chapman & Hall/CRC the R Series.

Wickham, H., M. Averick, J. Bryan, W. Chang, L. McGowan, R. François, G. Grolemund, et al. 2019. "Welcome to the Tidyverse." *Journal of Open Source Software* 4 (43): 1–6.

Wickham, H., and G. Grolemund. 2016. *R for Data Science: Import, Tidy, Transform, Visualize, and Model Data (https://oreil.ly/cBDs2)*. O'Reilly Media (*https://oreil.ly/ rzaZg*).

Wolfinger, R. 1993. "Covariance Structure Selection in General Mixed Models." *Communications in Statistics—Simulation and Computation* 22 (4): 1079–1106.

Wolpert, D. 1992. "Stacked Generalization." *Neural Networks* 5 (2): 241–59.

Wu, X., and Z. Zhou. 2017. "A Unified View of Multi-Label Performance Measures." In *International Conference on Machine Learning*, 3780–88. ACM.

Wundervald, B., A. Parnell, and K. Domijan. 2020. "Generalizing Gain Penalization for Feature Selection in Tree-Based Models" (*https://oreil.ly/R9wOm*).

Xie, Y. 2016. *bookdown: Authoring Books and Technical Documents with R Markdown* (*https://oreil.ly/CS7xl*). Boca Raton, FL: Chapman & Hall/CRC.

Xu, Q., and Y. Liang. 2001. "Monte Carlo Cross Validation." *Chemometrics and Intelligent Laboratory Systems* 56 (1): 1–11.

Yang, M., K. Kpalma, and J. Ronsin. 2008. "A Survey of Shape Feature Extraction Techniques" (*https://oreil.ly/dJLT3*). In *Pattern Recognition*, edited by P. Y. Yin. Rijeka: IntechOpen.

Yeo, I.-K., and R. Johnson. 2000. "A New Family of Power Transformations to Improve Normality or Symmetry." *Biometrika* 87 (4): 954–59.

Zeileis, A., C. Kleiber, and S. Jackman. 2008. "Regression Models for Count Data in R." (*https://oreil.ly/eIzza*) *Journal of Statistical Software* 27 (8): 1–25.

Zumel, Nina, and John Mount. 2019. "vtreat: A Data.frame Processor for Predictive Modeling" (*https://oreil.ly/v5yJS*).

Index

Symbols

!! operator, 207
!!! operator, 207
%>% (magrittr pipe operator), 21
+ operator for interaction specifications, 99

A

acceptance probability, simulated annealing, 227
acquisition functions, 218-222
activation function, 166
add_candidates(), 320
add_formula(), 81
add_model(), 83
add_role(), 106
add_variables(), 80-81, 83
Akaike information criterion (AIC), 338
aliasing, category values mapping to same hash value, 282
all_nominal_predictors(), 91
alternative hypothesis, 8
Alzheimer's study, and inferential model's limitations, 110
Ames housing data set, 45-52
    data analysis code development, 52, 61, 74, 87, 107, 147
    data budget, 56-58, 59
    encoding qualitative data in numeric format, 95-97, 273-283
    explainer algorithms, 286-297
    fitting models with parsnip, 66-73
    model tuning, 179-183
    in model workflow, 78-81, 84-87

performance evaluation, 125-128, 131-133, 135, 138-142, 145-146
recipes for feature engineering, 89-107
regression metrics, 112-113
resampling to compare models, 149-154
tuning specifications, 198-199
analysis data subset, training set, 128
analysis of variance (ANOVA) model, 10, 155-157, 158, 210, 338
analysis(), V-fold cross-validation, 131
apd_pca(), 315
apparent metric, 126
applicability, extrapolation measurement for predictors, 301, 307-315
applicable package, 313-315
arrange(), 18
assessment data subset, training set, 128
assessment(), V-fold cross-validation, 131
assumption-based predictive models, 9
autoplot()
    classification metrics, 117, 121
    comparing models with resampling, 151, 160, 163
    grids, 193, 195-196
    iterative search, 225
    model stacking, 321-323
    screening many models with workflows, 245
    simulated annealing, 231

B

bagged tree model, 269-271, 319
bake(), 260
Bayesian methods

model-based methods for variable importance, 286
model.matrix(), 82
modeldata package, 114
models and modeling, 3
    data spending, 55-61
    ensembles of models, 317-325
    explanation and prediction, 285-299
    fitting (see fitting model to data)
    performance evaluation (see performance evaluation)
    recipes for feature engineering, 89-107
    screening multiple models, 235-248
    stacking, 317-325
    terminology, 11
    tuning parameters (see tuning parameters)
    types of models, 6-10
    workflow for model (see workflows)
model_parts(), 292
model_profile(), 294
Monte Carlo cross-validation (MCCV), 133
multiclass classification metrics, 117-122
multilayer perceptron (MLP) model, 185-190, 208-210, 268-271, 318
multilevel data, spending data, 59
multivariate adaptive regression splines (MARS), 200, 319
mutate(), 239

N

naive Bayes, dimension reduction, 268-271
naming conflicts, tidymodels, 40-41
natural language processing, recipes for feature engineering, 103
natural splines, 100
Nelder–Mead simplex search method, 177
neural networks
    dimensionality reduction, 268-271
    emulation of nonlinear patterns, 180
    model stacking, 318, 324
    nonregular grids, 188-190
    racing methods, 208-210
    tuning parameters, 167, 174-175, 185-190
    validation sets for data evaluation, 59
nnet package, 186-186
nonregular grids, 188-190
normalize preprocessing method category, 343
ns(), 100
null hypothesis, 8

numeric outcome data, regression metrics, 112-113

O

objective argument, tune_bayes(), 222
one-hot encoding, 96
optimization, 167
    (see also tuning parameters)
    deciding on targets of, 168-172
    gradient descent methods, 167
    strategies for, 176-177
order(), 18
ordinal predictors (ordered qualitative column variables), 274
ordinary linear regression, 64
out-of-bag sample, 135
outcome variable, 11
    (see also classification models; regression models)
    inference models, 10
    model workflow, 76, 80
    prediction testing, 303-307
    probabilistic outcomes, 10, 118, 303-307
overfitting problem, tuning parameters, 173-175

P

p-values in inferential analysis, 157, 327, 332-334
parallel backend package, 143
parallel package, 143
parallel processing, 143-144, 200-206, 207, 208
parameters, 64
    (see also tuning parameters)
    estimating model, 64, 64, 173-175
    main arguments, 177
    penalty, 182, 193-196
param_info argument
    tune_bayes(), 222
    tune_grid(), 193
parsnip package, 63-74
    automatic submodel application, 200
    concrete mixture data example, 237
    creating model, 63-68
    extension packages, 73
    grid search, 186
    number of epochs/iterations for fitting, 186
    parameter arguments, 177
    poissonreg, 334

About the Authors

Max Kuhn is a software engineer at RStudio PBC. He is currently working on improving R's modeling capabilities. He was a Senior Director of Nonclinical Statistics at Pfizer Global R&D and applied models in the pharmaceutical and diagnostic industries for over 18 years. Max has a Ph.D. in biostatistics. He is the author of numerous R packages for machine learning and statistical modeling. He and Kjell Johnson wrote the book *Applied Predictive Modeling* (Springer), which won the Ziegel award from the American Statistical Association. A second book, *Feature Engineering and Selection* (CRC Press), was published in 2019.

Julia Silge is a data scientist and software engineer at RStudio PBC where she works on open source modeling tools. She holds a Ph.D. in astrophysics and has worked as a data scientist in tech and the nonprofit sector, as well as a technical advisory committee member for the US Bureau of Labor Statistics. She is an author, an international keynote speaker, and a real-world practitioner focusing on data analysis and machine learning. Julia loves text analysis, making beautiful charts, and communicating about technical topics with diverse audiences.

Colophon

The animal on the cover of *Tidy Modeling with R* is a European robin (*Erithacus rubecula*), native to continental Europe, the United Kingdom, western Russia, and northern Africa.

European robins are grey and brown, with a signature orange chest and white stomach. Males and females look similar, with the primary differentiator being the shape of the beak. They live around forests, plants, or trees. They don't necessarily migrate significant distances for winter (except for those living further north in Scandinavia and Russia), but females will move short distances away from males, who maintain the same territories in both winter and summer.

European robins typically begin breeding in March and will lay several clutches of 4 to 6 eggs each with an incubation period of 13 to 14 days. The clutches can overlap to a certain extent, with the male feeding the newborns while the female sits on the next clutch. The male bonds with the female by bringing her food, which can look like a mother feeding her young to an untrained observer. Their diet consists of insects, seeds, nuts, and fruits.

European robins have a very high mortality rate in the first year of life. After that, their life expectancy significantly increases. Ten percent of deaths after the first year are a result of fights between robins, as the males are very aggressive and protective of their territories. The orange chest gradually manifests itself in young robins; the initial lack of orange coloring reduces the chances of a fight over territory in their first year, when their chances of death are already so high.

Since 1960, the European robin has been the national bird of Britain, but it is less popular in other parts of Europe. They are a popular symbol of Christmas because Victorian mailmen were known as "redbreasts" due to their uniforms. The American robin is named for its similarity in appearance to the European robin, but they are not actually closely related.

The European robin's conservation status is Least Concern. Many of the animals on O'Reilly covers are endangered; all of them are important to the world.

The cover illustration is by Karen Montgomery, based on an antique line engraving from *Johnson's Natural History*. The cover fonts are Gilroy Semibold and Guardian Sans. The text font is Adobe Minion Pro; the heading font is Adobe Myriad Condensed; and the code font is Dalton Maag's Ubuntu Mono.

O'REILLY®

Learn from experts.
Become one yourself.

Books | Live online courses
Instant Answers | Virtual events
Videos | Interactive learning

Get started at oreilly.com.

©2022 O'Reilly Media, Inc. O'Reilly is a registered trademark of O'Reilly Media, Inc. | 175

Milton Keynes UK
Ingram Content Group UK Ltd.
UKHW051012111024
449497UK00007B/10